Lecture Notes in Economics and Mathematical Systems

Managing Editors: M. Beckmann and H. P. Künzi

Mathematical Economics

112

Jochen Wilhelm

Objectives and Multi-Objective Decision Making Under Uncertainty

Springer-Verlag
Berlin · Heidelberg · New York 1975

Author

Dr. J. Wilhelm
Institut für Gesellschafts- und Wirtschaftswissenschaften
Universität Bonn
Adenauerallee 24–42
53 Bonn/BRD

Library of Congress Cataloging in Publication Data

Wilhelm, Jochen, 1945-
Objectives and multi-objective decision making under uncertainty.

(Lecture notes in economics and mathematical systems ; 112) (Mathematical economics)
Bibliography: p.
Includes index.
1. Decision-making--Mathematical models. I. Title.
II. Series. III. Series: Mathematical economics.
HD69.D4W53 658.4'03 75-30514

AMS Subject Classifications (1970): 06A45, 28A05, 90A05, 90A10, 90C30

ISBN 3-540-07412-0 Springer-Verlag Berlin · Heidelberg · New York
ISBN 0-387-07412-0 Springer-Verlag New York · Heidelberg · Berlin

Printed in Germany
Offsetdruck: Julius Beltz, Hemsbach/Bergstr.

Contents

1. *The Concept of Objectives*

Managerial decisions as well as decisions in politics or daily life may be interpreted as results of choice-making processes; in this context, to decide means to choose amongst known opportunities of action - alternatives.- The idea of externally given selection opportunities being submitted to a subjective procedure of choice-making turns out to be true in a world of perfect information only, perfect regarding every real and possible dimension; here the decision maker is informed of all past , present, and future relevant events, all possible activities, and can describe them in every important detail.

If a decision is not to be formed in such a world - and this is in reality usually the case,- the reason may be a lack of information on the possible events on the one hand or a lack of information on the imaginable alternative actions on the other hand. This lack of information is partly of a systematic nature (i.e. basically unresolvable) and partly a consequence of incomplete information attainment. Whereas poor information on relevant events is usually of a systematic nature - economic decisions are typical decisions under uncertainty and risk respectively, or decisions where the responses of the remaining economic actors are unknown[1)] - the lack of information on possible alternatives is mainly a consequence of imperfect information attainment; furthermore, given information on whole classes of action opportunities is sometimes not taken into account - many entrepreneurs do not engage in lines of business other than their own lines, uninfluenced by more profitable opportunities.

Given the information level and given the initial state of the world, the decision situation is characterized a) by the set of alternatives, which the decision maker thinks are possibly faced with the given information b) by the set of responses of the environment, which may according to the decision maker's insight be generated by these actions, and c) by the information on actual realization or non-realization of these responses.

In a decision situation like that rational decisions become possible by pursuing an objective . But here arises another information problem: can it always be stated how far a certain final state of the world achieves the objective ? The classical decision theory makes such an assumption

1) Cf. the remarks made by Krümmel [1969, pp. 72-73].

on its basic model.[1] This supposition seems less realistic to us, especially as we consider an application of this basic model to problems of multi-period decisions, where information obtained in the course of time is integrated (flexible planning).[2] We will deal with this later on in more detail.

The following introduces a precise concept of what is meant by an "objective"; as the same objective may be pursued in different decision situations an exact description of the underlying decision situation is needed first.

1.1. The Decision Situation under Consideration

The classical decision theory used to start from the idea that the following formal structure is common to all real decision situations, and is sufficient as a basis of abstract decision theory: given a set A of actions there is a set Ω of possible responses of the environment; their interaction results in the set $A \times \Omega$ of possible final states; these final states are subject to a numerical evaluation carried out by the decision maker using a function $z : A \times \Omega \rightarrow R$, which assigns to each final state the related level of objective-achievement, in the following simply referred to as the *objective-achievement*. Describing decision situations in this way two problems prevent a definition of an objective: first, different objectives (probably) may generate the same function z with respect to the same decision situation; for another, if different decision situations are considered in the light of the parameters A_1, Ω_1, and z_1 or A_2, Ω_2, and z_2 respectively, it cannot be seen whether or not z_1 and z_2 describe the same objective. The reason is that in this model only the objective-achievement of a final state is regarded as a property of that state, ignoring the process of working up and evaluating information on real properties of that final state. The following will attempt to describe properties of final states and, based on such a description, to develop a model of information processing on which the construction of objectives is heavily based.

1) Schneeweiß [1967, p. 8], Radner [1964, p. 179].
2) Cf. Riemenschnitter [1972, p. 72] especially for the problems of so-called "dead knots" ("tote Knoten") within decision trees.

Throughout, we will take our standpoint from the decision situations with given information levels; hence we will not be concerned with problems of information attainment, information evaluation, etc.[1] These decision situations may be described by four components:

1) the initial state of the world
2) the set of possible alternatives (actions) corresponding to this initial state
3) the set of possible responses of the environment
4) the information on realization or non-realization of these responses.

ad 1. The economic and social environment as well as the existing technologies and the special initial endowment - the decision maker's initial resources - belong to the initial state of the world. As we will see, there is no need to describe this initial state in full generality; there must only be known decision-relevant aspects of the final state, which is generated by the interaction of the initial state concerned with a certain action and a response of the environment.

ad 2. The alternatives need not be characterized in more detail either; similar remarks hold for them. The concept of an alternative stands for single actions corresponding to a point or a partial period of the planning period ("stiff" planning) as well as for strategies which prescribe such actions at each point or partial period respectively, depending on information (on realized responses of the environment) obtained at those points or partial periods (flexible planning). We will assume the alternatives to incorporate the initial state of the world they start from. This has no substantial consequences, only formally simplifying ones.

ad 3 and 4. Regarding the description of the environmental responses we may refer to the preceding remarks. We further assume the information provided by 3 and 4 to be given in the following way: let A be the set of imaginable alternatives corresponding to the initial state of the world and to the decision maker's information level. For each $a \varepsilon A$ let $(\Omega_a, \mathcal{A}_a)$ be a measurable space[2] where Ω_a is the *set of environmental responses to action* a. The set of possible final states resulting from an action a may be characterized by $\{a\} \times \Omega_a$. The information of 4 is assumed to be given as follows:

1) For these problems we refer to Albach [1969, pp. 720-727] and the references given there.
2) For a definition of a measurable space and for all probability theoretic material we refer to Neveu [1969], here p. 27.

let $M(A_a)$ be the set of all positive and σ-additive measures ψ on A_a with $\psi(\Omega_a) = 1$ and $\psi(\emptyset) = 0$, i.e. the set of probability measures on (Ω_a, A_a). The information of 4 is represented by a mapping $\mu : a \to \mu_a \in M(A_a)$; this map assigns to each action a the probability measure of the resulting environmental responses and thus of the arising final states.

This assumption turns out to be quite general for it comprises risk, uncertainty - as far as subjective probabilities seem reasonable - and, at least partially, the game situation.[1)] We will not enter the discussion of the nature of probabilities;[2)] it may be left to the reader's attitudes which meaning is given to the probability measures μ_a. Following our preceding arguments the decision situation may be described by a set A together with a class $\{(\Omega_a, A_a, \mu_a) | a \in A\}$[3)4)] of probability spaces.

In this representation final states are objects without attributes. The description, however, of the decision situation needs, as indicated before, to reflect the information on the relevant attributes of the final states available to the decision maker. Attributes establish relations between final states - regardless of whether or not they belong to the same decision situation - along the following line: final states correspond to each other by such a relation if they cannot be destinguished with respect to the attribute concerned. The specification of this relation certainly does not completely cover the attribute in question, but it is an important characteristic common to all attributes. Since we cannot cover the manifold additional characteristics uncommon to all attributes, we confine ourselves to the aforementioned relation.

Relations of the aforementioned kind generated by attributes of final states, may be of different types; they are characterized by the environmental data of a decision situation and by the information produced by processing these data which generates information of "higher quality" or "derived information". This processing is done by the decision maker

1) This classification of decision situations is due to Knight [1921, p. 233].
2) Cf. Keynes [1950, pp. 13-14], Carnap and Stegmüller [1959]. Axiomatical treatments of these problems may be found as a context of developing the expected-utility-hypothesis in Savage [1954], Pfanzagl [1968, pp. 195-196] and in Rabusseau and Reich [1972].
3) In the following this class will throughout be denoted by $\langle \Omega, A, \mu \rangle_A$.
4) The basic model of decision theory assumes $\Omega_a = \Omega$, $A_a = A$, $\mu_a = \mu$ for all $a \in A$, cf. Schneeweiß [1967, p. 12].

in view of the chosen objective. Some examples may be given as follows: final states obtaining the same amount of profit (however defined), final states generated by the employment of the same production factors, final states causing the same level of environmental pollution (however defined) and final states characterized by the same selling price for a certain product.

These relations will be built into our concept in two steps: first we will deal with the environmental data of a decision situation, secondly we will establish relations between data belonging to different situations; establishing these relations is combined with a processing of data to information of higher quality carried out with regard to aspects of objective-relevance. As we see, in the second step other characteristics of attributes than the above-mentioned may implicitely be taken into consideration.

Take a given decision situation $(A,<\Omega,\mathcal{A},\mu>_A)$; the set of final states $E:=\bigcup_{a\varepsilon A}\{a\}\times\Omega_a$ is characterized by the environmental data. Data terms are given such as *price, product quality, net cash receipts, stock of factors* etc.. Let N be the set of names which are necessary for the description. The above-mentioned relations can now be represented by *named* equivalence relations.[1)]

Definition 1.1. By an *aspect* of the final states E we mean an ordered pair (n,Q) consisting of a name $n \varepsilon N$ and an equivalence relation Q on the set E.

As easily seen different aspects may induce the same relation on the set of final states of a decision situation. If $[E]$ denotes the set of all equivalence relations on E, a set $G \subset [E]\times N$ of aspects is technologically given by the environmental data of the considered situation; G meets the requirement: $(n,Q)\ \varepsilon\ G$ and $(n,Q')\ \varepsilon\ G \Rightarrow Q = Q'$, i.e. G is the graph of a mapping (we denote the graph and mapping by the same symbol) $G : N \rightarrow E$. Naturally G need not to be known completely; for one thing that is unnecessary because many aspects are by no means relevant for certain objectives; for another that is in fact not true, for, in gene-

1) By a *binary relation* on a set M is understood a subset Q of M×M. An *equivalence relation* on the set M is a binary relation Q on M with the properties:
(i) for all $x\varepsilon M$ we have $(x,x)\varepsilon Q$ (reflexivity)
(ii) for all $x,y\varepsilon M$ we have $(x,y)\varepsilon Q \Rightarrow (y,x)\varepsilon Q$ (in the notation of relations $Q \subset Q^{-1}$ (symmetry))
(iii) for all $x,y,z\varepsilon M$ we have: $(x,y)\varepsilon Q$ and $(y,z)\varepsilon Q \Rightarrow (x,z)\varepsilon Q$ (in the notation of relations $Q\cdot Q \subset Q$ (transitivity))
For a binary relation we often write xQy instead of $(x,y)\varepsilon Q$.

ral not all information on a not yet realized state of the world are known. In discussing our concept of objectives under uncertainty we will deal with this in more detail.

For every aspect (n,Q) we may construct the set $E_{/Q}$ of equvivalence classes with respect to Q.[1] Let s_Q be the corresponding canonical mapping. By forming the cartesian product[2] $\prod_{n\varepsilon N} E_{/G(n)}$ the final states become *multi-dimensional objects* by virtue of the mapping $\prod_{n\varepsilon N} s_{G(n)}$ [3] where every aspect corresponds to one dimension.

The first step, the explicit description of attributes of final states has worked well: every aspect quotes which final states have the same property with respect to this aspect; e.g. the aspect, *selling price for product* P, reflects which final states are characterized by the same selling price for product P; an equivalence class with respect to this aspect collects all final states which show the same price of, e.g. 5 DM per unit of P. The depicted equivalence class may then be identified with the real number 5. Considering a second aspect, *output of product* P, allows the related equivalence classes to be described by real numbers which quote the amount of output connected with the corresponding final states. Accordingly, a final state characterized by the price 5 DM and the output 30 000 units may be interpreted as the 2-dimensional object (5 , 30 000); for every additional aspect a further dimension would be added correspondingly.

For sake of an easier understanding we will carry out the second step within the definition of an objective under certainty first.

1.2. *The Concept of Objectives under Certainty*

1.2.1. *'Aspects' and 'Points of View'*

In our context a *decision situation under certainty* means the following:

- for all $a \varepsilon A$ the set Ω_a consists of one and only one

1) Given an equivalence relation Q on the set M the *quotient set of* M *by (with respect to* or *modulo)* Q will be denoted by $M_{/Q}$. It is defined as follows: for each $m\varepsilon M$ let be $Q_m := \{y\varepsilon M \mid (m,y)\varepsilon M\}$; then $M_{/Q}$ is defined by the equation $M_{/Q} := \{Q_m \mid m\varepsilon M\}$. Hence the quotient set induces a partition of M into *equivalence classes*. The surjective mapping $m \rightarrow Q_m$ is usually called the *canonical projection* or the *canonical mapping*.
2) The definition of a cartesian product is given in footnote 1 of page 8 of these notes.
3) For this terminology we refer to Klahr [1969, p. 596]

element. Hence the set of final states is one-to-one to the set A of actions; accordingly we identify both sets formally.

- G is completely known.

An objective is intuitively understood to mean a certain method of ordering which in different situations helps to a ranking arrangement of alternatives with respect to the same points of view in a consistent way; this arrangement is made by an evaluation according to various criteria, which prepare the given information material. The operation of pursueing an objective is, as a rule, rather linked to certain specific situations than to whole classes of imaginable decision situations. Thus, in order to define an objective we take as a basis an arbitrary class $\mathfrak{A}$ of decision situations. Thereby we will assume to be given a set of names for the aspects of the environmental data. As we have seen above, $\mathfrak{A}$ consists of activity sets; to each A there corresponds a set G_A, the set of aspects of A uniquely determined by the environmental data of the situation concerned.

By $\mathfrak{A}^*$ we denote the coproduct[1)] $\coprod_{A\in\mathfrak{A}} A$; it comprises all possible actions (within $\mathfrak{A}$) or final states respectively in such a way that to each action $a \in \mathfrak{A}^*$ uniquely corresponds a decision situation A with $a \in A$. Instead of $inj_A\, A$ we write A^*. Requiring an objective to be pursued in different situations consistently presupposes an idea of homogeneity of final states resulting from different situations. In order to develop such an idea we use a method similar to that one used in introducing aspects. By aspects we described homogeneity of actions resulting from the *same* decision situation in view of the basic data. Homogeneity with respect to aspects is a material one concerning the environmental data. Homogeneity, however, with respect to the *points of view* here, to define is to describe homogeneity with respect to certain criteria which are chosen by the decision maker in accordance to the objective under con-

1) Let $\{X_i | i \in I\}$ be a class of sets whose union is $E = \bigcup_{i\in I} X_i$. The *coproduct* of this class is defined by $\coprod_{i\in I} X_i := \{(i,x)\in I\times E | x\in X_i\}$. By $inj_i : X_i \rightarrow \coprod_{j\in I} X_j$ we denote the *canonical injections:* $inj_i(x)=(i,x)$ for all $i\in I$ and all $x\in X_i$. (Bourbaki [1970, E.R. 18])
The *product* (or *cartesian product*) of this class is defined by $\prod_{i\in I} X_i = \{\psi : I \rightarrow E | \psi(i)\in X_i$ for all $i\in I\}$. By the *canonical projections* we understand the mappings $pr_i : \prod_{j\in I} X_j \rightarrow X_i$ with $pr_i(\psi)=\psi(i)$ for all $i\in I$. (Bourbaki [1970, E.R. 20])! For products and coproducts as parts of the category theory we refer to Lang [1965, p. 33].

sideration; additionally they serve for a condensation and conversion of the basical data described by the aspects; points of view combine different aspects in a sense still to be made precise; e.g. the point of view *value of the output* combines the aspects *output* and *selling price* by multiplication.

For a definition of the term *point of view* we start from the representation of actions as multi-dimensional objects, since points of view are only to be related to those properties of actions which are represented by aspects. If $A \in \mathfrak{A}$ holds and if G_A is the related set of aspects and N_A the associated set of names ($N_A = pr_1 G_A$), then for each $n \in N_A$ the relation $G_A(n)$ an equivalence relation $G_A(n)$ on A^* in a natural way; there exists an i such that the following diagram becomes commutative:

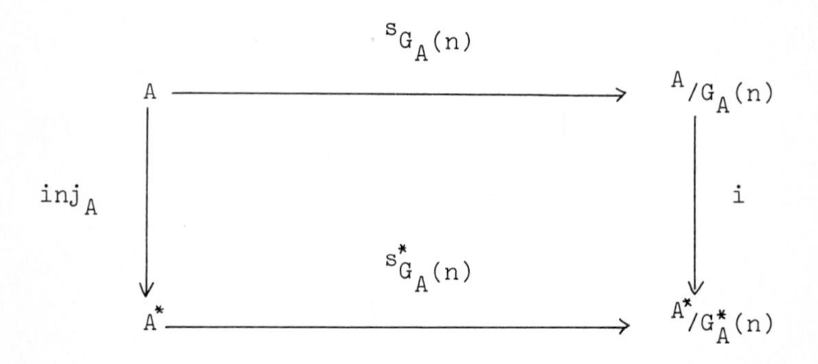

The product $\bar{A} := \prod_{n \in N_A} {}^{A^*}/G^*_A(n)$ completely corresponds to the representation of A as a set of multi-dimensional objects; but unlike that representation, and that is what we need, $\bar{A}$ reflects the original decision situation the objects belong to: we have $\bar{A} \cap \bar{B} = \emptyset$ as long as $A \neq B$. Let $\bar{\mathfrak{A}}$ be the union of all $\bar{A}$ where A runs through $\mathfrak{A}$; $\bar{\mathfrak{A}}$ will be called the *data-dimension space* of $\mathfrak{A}$. Each action a and each $A \in \mathfrak{A}$ with $a \in A$ generate one and only one representation of a as a multi-dimensional object in $\bar{\mathfrak{A}}$ by virtue of the map

$$\prod_{n \in N_A} s^*_{G_A(n)} \cdot inj_A .$$

A natural representation map may be defined as follows:

$$\Delta : \mathfrak{A}^* \to \bar{\mathfrak{A}}$$

1) Let $f: X \to Y$ be an arbitrary mapping. Let $Q \subset Y \times Y$ be an equivalence relation. Then $Q_f := \{(x,z) \in X \times X \mid (f(x), f(z)) \in Q\}$ is an equivalence relation on X. This is easily seen by the reader.
Take here the map inj_A^{-1} for f.

$$\Delta(A,a) = \prod_{n \varepsilon N_A} s^*_{G_A(n)}(A,a)$$

To each action Δ assigns the related representation as a multi-dimensional object according to the decision situation the action belongs to.

Definition 1.2. By a *point of view* for the class $\mathcal{A}$ we mean a pair (n,Q) consisting of a name n and an equivalence relation Q on the data-dimension space of $\mathcal{A}$, i.e. $Q \varepsilon [\bar{\mathcal{A}}]$.[1)] By Δ the relation Q induces an equivalence relation Q_Δ on $\mathcal{A}^*$ too. Given the set $\bar{N}$ of names the set of all possible points of view is denoted by PV($\mathcal{A}$). It may be assumed that $\bar{N}$ is a countable set, hence PV($\mathcal{A}$) may be identified with $\coprod_{n \varepsilon \bar{N}} X_n$ where X_n is the set of all equivalence relations on the data-dimension space of $\mathcal{A}$ whichever the chosen n is. The following is based on this assumption.

By the concept of an aspect the final states have been defined as multi-dimensional objects; a point of view defines a (equivalence-)relationship on the underlying space of dimensions, the data-dimension space. Thus, by linking the final states with their representations in this space the point of view defines a relation between the final states too, whichever may be the decision situation they belong to. Accordingly, the point of view processes the information given by the images of final states in the data-dimension space to new information of higher quality.

A simple example might illustrate the developed ideas.

Let a set of actions be described as follows: let n,m be some integers; n completely divisible goods can be produced using m completely divisible input factors, and can be sold in any amount. In order to produce one unit of good i we require p_{ij} units of factor j. There are K_j units of factor j available (i=1,...,n; j=1,...,m).

$$A(n,m,P,K) := \{x \varepsilon R^n \mid x \geq 0 \text{ and } P.x^t \leq K\}$$

describes the activity set of a decision situation using the parameters $n,m,P=(p_{ij}),K=(K_j)$ and incorporating the initial state of the world.

We now denote by c_j the costs of consumption of one unit of factor j (j=1,...,m) and by p_i the selling price of good i (i=1,...,n). These two pa-

1) Hence, regarding points of view the following holds like before: different points of view may establish the same equivalence relation on the data-dimension space of $\mathcal{A}$. Nevertheless in cases no confusion may arise we identify points of view and their corresponding equivalence relations.

rameters are the characteristics of the environmental responses. Hence the arising final states of a decision situation are determined by the quantities n,m,P,K,c,p. We first introduce the aspects:

1) *number of goods* (n)
2) *number of factors* (m)
3) *technology of production* (P)
4) *initial resources* (K)
5) *factor prices* (c)
6) *selling prices of goods* (p)

By these aspects one and the same equivalence relation is induced on the set of final states of a decision situation: the only equivalence class is the set itself; the chosen actions do not have any influence.

7) *number of units of product i produced* (x_i)
8) *remaining stock of factor j* $(K_j - \sum_i p_{ji}x_i)$

At every decision situation described in that way final states and production vectors x may be identified. The aspects 7 are given by the relations $xQ_iy \iff x_i=y_i$; the aspects 8 induce the relations $x\bar{Q}_jy \iff$ $\iff \sum_i p_{ji}x_i = \sum_i p_{ji}y_i$. Naturally there may exist some more aspects of the mentioned situation, but the given selection is sufficient for our purposes.

We now introduce some points of view: denoting the remaining stock of factor j by r_j the *dimensions* of a final state are given by the vector (n,m,P,K,c,p,x,r).

1) *costs of production* (C)
2) *sales revenue* (S)

where $C(n,m,P,K,c,p,x,r) = \sum_j (c_j \cdot \sum_i p_{ji}x_i)$ and $S(n,m,P,K,c,p,x,r) = \sum_i p_ix_i$ are the functions which define the corresponding relations on the dimensions of final states.

1.2.2. *The Category of Ordered Topological Spaces. The Ordinal and the Cardinal Category*

We need some mathematical constructions for our subsequent arguments.

By $|G|$ we denote the class of non-empty totally ordered sets of a universe provided with a topology finer than the related interval topology.[1] Take $G, G' \varepsilon |G|$: by $Mor_G(G,G')$ we denote the set of continuous order-homomorphisms[2] from G into G'. $(|G|, Mor_G)$ becomes a category, denoted by *G*, the *category of ordered topological spaces*.

From *G* we construct another category denoted by *O*:

The objects of *O* are pairs (G,g) consisting of an object G of *G* and of a morphism $g \varepsilon Mor_G(G,R)$ where R is provided with its usual topology and g is injective considered as a mapping.

The morphisms of *O* are as follows: given two objects (G,g) and (G',g') such a morphism is a pair (h_1,h_2) of *G*-morphisms where $h_1 \varepsilon Mor_G(G,G')$ and $h_2 \varepsilon Mor$ (Im g,Im g') and where the following diagram is commutative:

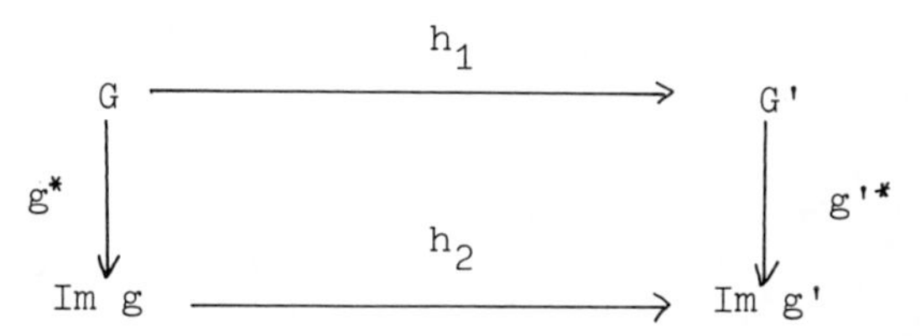

Considering Im g and Im g' provided with their respective topologies as objects of *G* and denoting the corresponding inclusions by inc and inc' respectively (treated as morphisms of *G*), the morphisms g* and g'* are uniquely determined by the following commutative diagrams:

---

1) For the notion of a *universe* we refer to Schubert [1970, p. 16]. For all topological notions not defined here or in the following and for all statements on topological material not proved here or in the following see Bourbaki [1965].
For the notion of the *interval topology* we refer to definition 1.8. to be found in our excursus on the Bernoulli-principle as well as to Pfanzagl [1968, p. 61]. For an arbitrary ordered set M, by $\mathcal{J}_M$ we denote the corresponding interval topology.
2) Order-homomorphisms are mappings $g : G \rightarrow G'$ with the property: for all $a,b \varepsilon G$ we have $a \leq b \Rightarrow g(a) \leq g(b)$; cf. definition 1.11.
3) For the concept of a category cf. Schubert [1970] or Lang [1965, p. 25].

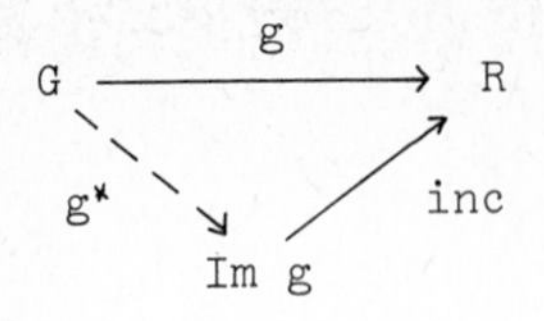

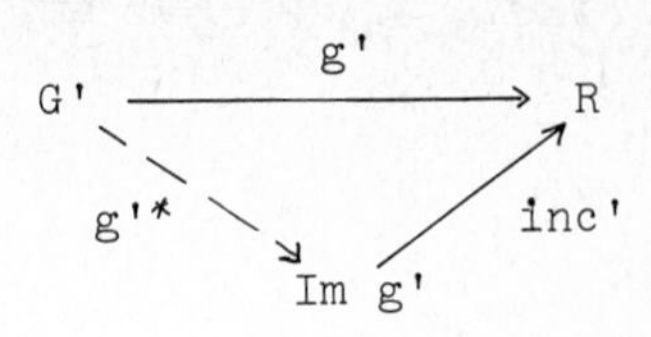

$\mathcal{O}$ will be called the *ordinal category*. It consists of the representations of continuous orderings by continuous utility functions.[1] By the *cardinal category* we mean the subcategory K of $\mathcal{O}$ whose collection of objects is the same as that of $\mathcal{O}$ and whose sets of morphisms are subsets of the sets of $\mathcal{O}$-morphisms with the property:

> For all objects $P,P' \varepsilon |K| = |\mathcal{O}|$ we have: $(h_1,h_2) \varepsilon Mor_K(P,P')$ if and only if $(h_1,h_2) \varepsilon Mor_{\mathcal{O}}(P,P')$ and h_2 may (as a mapping) be extended to all of R such that the extension becomes a semi-positive[2] linear transformation.

By the *nominally ordering category* we mean a subcategory N of the ordinal category which is defined as K if "semi-positive" is replaced by "identity- or 0-mapping". More generally, let $L_+(R)$ be the group of positive linear transformations of R; given a subgroup D of $L_+(R)$ by $\bar{D}$ we denote the union of D and the set consisting of the 0-mapping. By a D-*-category* we mean a subcategory of the ordinal category defined as K, if "semi-positive" is replaced by "a member of $\bar{D}$". By this definition the cardinal category is the $L_+(R)$-category, the nominally ordering category is the $\{id\}$-category; the *ratio-category* is a D-category where D is the group of dilations: $D = \{\psi \varepsilon L_+(R) | \psi(0)=0, \text{and } x>0 \Rightarrow \psi(x)>0\}$.

1.2.3. *The Definition of an Objective under Certainty*

An objective is to meet two requirements

- an objective must allow the statement, whether or not one of two

1) For the notion of a utility function is refered to Debreu [1959, p.55]. Regarding the terms *ordinal* and *cardinal* see Pfanzagl [1968, p. 74]. In our context these terms above all have respect to the isomorphisms of the category in question. Utility functions are special faithful order-homomorphisms (cf. def. 1.11.); continuous utility functions exist, if and only if the topology of G is separable (Pfanzagl [1968, p. 75], 4.2.3. Theorem).
2) By a *semi-positive linear transformation* is to be understood a mapping $\psi : R \to R$ such that $\psi(x) = ax + b$ holds for all $x \varepsilon R$ where $a \geq 0$ and $b \varepsilon R$, and furthermore $a = 0$ implies $b = 0$. If $a>0$ holds we will simply speak of a *positive linear transformation*.

arbitrarily given final states is more desirable with respect to the objective;

- this statement has to be made for different decision situations according to the same criteria.

Speaking of the objective *profit-maximization* requires a definition of profit which has to be independent of the specific decision situation. In continueing the above-mentioned example such a definition may be given by the expression "sales revenue ./. costs". Hence the objective can be defined on the basis of selected points of view. But in so doing the problem arises that alternate points of view may be used in defining profit as well without imputing a different meaning to the objective: if, within the given example, *value of the initial resources* $\sum_j c_j K_j$, *sales revenue* and *value of the remaining stock of factors* $\sum_j c_j r_j$ are chosen as points of view, profit may also be defined by the term "sales revenue ./. value of initial resources + value of the remaining stock of factors". Both definitions, though based on different sets of points of view, lead to the same objective. Hence, defining the notion of an objective it must be stated whether or not certain sets of points of view are equivalent with respect to the objective concerned.

A second problem is related to the fact that, e.g., the expression e^p, where p symbolizes the profit, induces the same ranking of final states as the profit itself, since the exponential function is monotone increasing; nevertheless this expression will in general not be accepted as a plausible definiton of profit at all. Hence, defining the notion of an objective it must be stated whether or not certain numerical representations of the ranking induced by the objective concerned are feasible.

These remarks made we are prepared to construct the abstract notion of an objective, having the mentioned requirements and problems in mind. In a first step we give a precise meaning to the terms *more desirable* or *less prefered*, which are to be used in describing the ranking order induced by an objective.

Definition 1.3. By a complete *preference relation (or preference ordering)* on an arbitrary set M we mean a binary relation Q on M meeting the following requirements:

(i) $x,y \varepsilon M \Rightarrow (x,y) \varepsilon Q$ or $(y,x) \varepsilon Q$ *(completeness)*

(ii) $x \varepsilon M \Rightarrow (x,x) \varepsilon Q$ *(reflexivity)*

(iii) $(x,y) \varepsilon Q$ and $(y,z) \varepsilon Q \Rightarrow (x,z) \varepsilon Q$ *(transitivity)*

The relation $Q' = Q \cap Q^{-1}$ is an equivalence relation on M, the so-called *indifference relation* associated with Q. The quotient set $M_{/Q'}$ is totally ordered (in addition to (i), (ii) and (iii) we have: $(x,y) \varepsilon Q$ and $(y,x) \varepsilon Q \Rightarrow x=y$ with respect to the preference relation induced on this quotient set) in a natural way. If we have $(x,y) \varepsilon Q$ we say y *is not less prefered than* x or x *is not more desirable than* y; if $(x,y) \varepsilon Q'$ holds we say x *is indifferent to* y.

In a second step we try to meet the requirement that the preference statements are to be made in all decision situations along the same criteria; actions or final states respectively, not differing with regard to points of view relevant for the objective, should not be treated differently. Given a point of view $\gamma \varepsilon PV(\mathfrak{A})$ by $\mathfrak{A}_\gamma := \bar{\mathfrak{A}}_{/\gamma}$ the equivalence classes modulo γ are given. By ρ_γ we denote the associated canonical projection. Let ψ_γ given by $\psi_\gamma = \rho_\gamma \cdot \Delta : \mathfrak{A}^* \to \bar{\mathfrak{A}}_{/\gamma}$ (notation as in definition 1.2.). Hence ψ_γ assigns to each final state (A,a) the related equivalence class modulo γ. The mapping ψ_γ induces on $\mathfrak{A}^*$ just the relation Q_Δ given in definition 1.2.. Final states (A,a) and (A',a') with $\psi_\gamma(A,a) = \psi_\gamma(A',a')$ do not differ with regard to the point of view γ. Hence preference statements will be applied to the images of mappings like ψ_γ only.

Both problems - the possible equivalence of different sets of points of view with respect to the definition of an objective, and the feasibility of different numerical representations of the same preference ordering - will be dealt with in the third step: the definition of the term "objective".

Definition 1.4. Let $\mathfrak{A}$ be a class of decision situations under certainty and *V* be a subcategory of *O*. By a *V-objective* we mean an equivalence class (the associated relation will be defined below) of the tripel (H,Q,g) where the following requirements are met:

(i) H is a finite set of points of view for $\mathfrak{A}$, $H \subset PV(\mathfrak{A})$.

(ii) Q is a complete preference relation on the set $E_H := \prod_{\gamma \varepsilon H} \mathfrak{A}_\gamma$.

(iii) g is a mapping $g : E_H \to R$.

(iv) The pair (E_{HQ}, g) is an object of the category *V* where $E_{HQ} := E_{H/Q'}$; especially, g is a faithful order-homomorphism, and g is continuous with respect to the interval topology.

Two tripels (H_1,Q_1,g_1) and (H_2,Q_2,g_2) fulfilling (i) - (iv) will be called *equivalent* if we have:

(v) There exists a V-isomorphism $(h_1,h_2) : (E_{HQ_1},g_1) \to (E_{HQ_2},g_2)$ such that the diagram

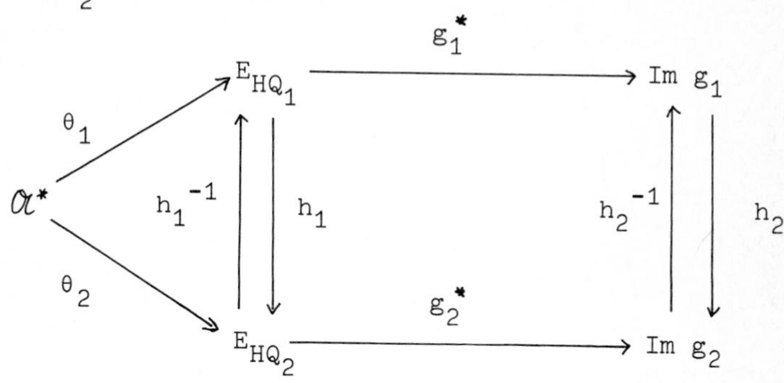

is commutative, where $\theta_i := \pi_i \cdot \prod_{\gamma \varepsilon H_i} \psi_\gamma$ and π_i is the canonical projection associated with $Q_i^!$ (i=1,2) and the sign * is to be understood as on pp. 11-12.

The set H is called the *set of objective-relevant points of view*. The set E_H is called the *set of consequences*[1], the mapping $\psi_H := \Pi\psi_\gamma$ is the *result function*, the mapping $g\cdot\pi\cdot\psi_H$ is the *objective function* associated with the objective. The mapping $g\cdot\pi$, the *utility function*, represents the decision maker's evaluation of the consequences. The mapping g is called the *reduced utility function*, the set E_{HQ} *reduced set of consequences*, and the mapping $\pi\cdot\psi_H$ *reduced result function*. An element (H,Q,g) of the equivalence class of an objective will be called a *representation* of the objective; frequently we shall identify representations and objectives with each other.[2]

By the point (i) the definition establishes the set of objective-relevant points of view; by (ii) the preference structure defined by the objective

1) An element of the set of consequences may be understood as another representation of an action as a multi-dimensional object: but these dimensions are of a more subjective nature, for they are with respect to an individual objective.
2) It is a remarkable fact that the terms E_H, ψ_H, $g\cdot\pi$, E_{HQ} and $\pi\cdot\psi_H$ heavily depend on the chosen representation, whereas the objective function does not.

is described, by (iii) a certain numerical representation of this structure is given, and by (iv) the class of numerical representations is described which the chosen one belongs to. Condition (v) simultaneously deals with both of the previous-mentioned problems: alternate choices of points of view and of numerical representations lead to the same objective, if the generated objects are isomorphic where the related isomorphisms have to be contained in a certain set of morphisms which is characteristic to the objective.

Proposition 1.1. Let (H,Q_1,g_1) and (H,Q_2,g_2) represent the same objective; then we have the equation:

$$Q_1 \cap (\mathrm{Im}\ \psi_H \times \mathrm{Im}\ \psi_H) = Q_2 \cap (\mathrm{Im}\ \psi_H \times \mathrm{Im}\ \psi_H).$$

Proof: Let $(x,y)\varepsilon Q_1$ and $x,y\varepsilon \mathrm{Im}\ \psi_H$, i.e. $x=\psi_H(a)$ and $y=\psi_H(b)$. Let (h_1,h_2) be the isomorphism of (v) in def. 1.4. h_1 is an order-homomorphism from E_{HQ_1} into E_{HQ_2}. We have: $h_1 \cdot \pi_1 \cdot \psi_H = \pi_2 \cdot \psi_H$. If $\bar{Q}_1$ is the order on E_{HQ_1} induced by Q_1, we have $(\pi_1(x),\pi_1(y))\varepsilon\bar{Q}_1$. Since h_1 is an order-homomorphism, $(h_1 \cdot \pi_1(x), h_1 \cdot \pi_1(y))\varepsilon\bar{Q}_2$ holds where $\bar{Q}_2$ is induced by Q_2 on E_{HQ_2}. From $h_1 \cdot \pi_1(x) = h_1 \cdot \pi_1 \cdot \psi_H(a) = \pi_2 \cdot \psi_H(a) = \pi_2(x)$ and (analogously) $h_1 \cdot \pi_1(y) = \pi_2(y)$ we have $(\pi_2(x),\pi_2(y))\varepsilon\bar{Q}_2$ and hence by definition $(x,y)\varepsilon Q_2$. The converse can be shown analogously.¶

Proposition 1.1. states that the preference relation Q is in some sense uniquely determined by the objective as soon as a certain set of objective-relevant points of view is chosen.

In many practical cases the set of consequences is measured by numerical terms. It is easily shown that in such cases the set of consequences may be substituted by a subset of the real numbers, the reduced utility function by a "numerical" utility function defined on that subset.

Immediately seen is the fact that the objective function $z=g \cdot \pi \cdot \psi_H$ of a V-objective induces an ordering on the set $\mathfrak{A}^*_{/z}$ [1] such that the objective uniquely defines an object $P:=(\mathfrak{A}^*_{/z},z)\varepsilon|V|$ up to isomorphisms of the form (id,h_2); hence the objective function of a V-objective is unique only up to certain - on the choice of V dependent - order-preserving functions from R to R.

Remark 1. The concept of utility functions has prevailingly been introduced with regard to the treatment of situations under uncertainty; in

---

1) $\mathfrak{A}^*_{/z}$ is an abbreviation for the quotient set which is generated by z and the equality relation in R (cf. p. 8, footnote 1).

such situations first this concept becomes really powerful. Under certainty weaker concepts would work well. But the utility-function approach does not restrict us too much if we assume $V = 0$ to hold; for, under weak assumptions a preference relation may be represented by a utility function unique up to continuous and monotone transformations in R.[1]) The goal of a decision-making process, to ascertain "best" elements (actions), may be attained under quite weak conditions in case of certainty.[2]) But a computational treatment of these problems usually needs a numerical objective function.

Remark 2. The concepts of reduced sets and functions have been introduced for formal reasons only; empirical relevance is only due to the non-reduced terms.

1.2.4. An Illustration of the Introduced Concepts

As we have seen, points of view may be described by equivalence relations on the data-dimension space of the underlying class of decision situations. In the previously given example the dimensions of a final state were described by an 8-tupel like (n,m,P,K,c,x,r). Choosing for H the set consisting of the points of view defined on p. 10 H may be represented by the depicted two functions C (cost function) and S (sales-revenue function). If "profit maximization" is the objective to be pursued the preference relation Q may be represented by the function Pr:= S - C. It is seen that Q depends heavily on the set H of objective-relevant points of view. According to which units of measurement will be feasible the category V will be chosen. If the nominally ordering category is taken for V the function Pr is uniquely determined by H and Q and is identical with the utility function of the objective; at the same time we got an example of a numerically measured set of consequences associated with an objective.

1) Cf. p. 12, footnote 1. For an example take the utility function of a consumer, Debreu [1959, pp. 55-57].
2) Cf. corollary of theorem A in the appendix.

1.2.5. *Side Conditions under Certainty*

Up to now, formulating the decision problem has assumed that the set of considered and realizable actions has already been marked off and given. This is in fact the result of some kind of information processing. It cannot be stated whether or not an action is actually realizable as long as the combination of some information has not been done on the initial state of the world and on the responses of the environment to the hypothetically selected action. In discussing this question the instrument of side conditions is useful. Side conditions condensate a bundle of informations to a (0,1)-statement on actions. Usually side conditions are not only constructed to mark off realizable actions from unrealizable ones, but also to mark off desired from undesirable actions.[1)]

Essentially information of the following nature is used to restrict the feasibility of alternatives when making decisions on a firm's business:

a) *technical relations:* we already adduced capacity-restrictions as an example of side conditions for decisions on production.

b) *economic environment:* general economic environment like state of business, central bank's policy, wage policy; individual economic environment like state of factor markets and markets for sale of the firm's goods, whichever is concerned.

c) *legal and social standards:* Bankruptcy Code, law of labour, law of contract, safety and pollution of the firm's plants at certain minimum or maximum levels respectively.

d) *subjective desirabilities:* maintaining a certain market position, observing a constant level of dividends, holding a minimal amount of liquid assets. These subjective desirabilities may have different sources

- they may be generated by the decision maker's aspiration;
- they may result from the formal and informal interaction of several decision making units (e.g. supervisory board, managing board, general meeting of a company in AG form);
- they are to avoid certain reactions of the environment (reactions

1) Heinen [1966] has made similar remarks in differing between so-called side conditions of type A and type B. This specification is of some importance to the problem of the so-called "goal-programming" also: here the problem is discussed that by setting "goals" (= side conditions of type B) the set of feasible actions may become empty. A first approach may be found in Balderstone [1960]. For further references we refer to p. 63 of these notes.

of the small shareholders, of the legislator, of the Press etc.

Generally a) and b) imply side conditions of type A: they exclude unrealizable actions. c) and d) lead to conditions which cut off undesirable (socially undesirable and partially under threat of punishment, or individually undesirable) actions. In most cases sanctions for violating a condition mentioned in c) have consequences so unpleasant that their neglect will not be considered. Hence, with good reasons conditions of this kind may be treated as side conditions of type A. Proper side conditions of type B are consequences of d).

We will deal with side conditions in terms of our concept of objectives. Replacing the sets of feasible actions by the sets of "considered" actions (the statement whether or not an action is feasible can be made after having processed the relevant information on considered actions first) and extending the sets of aspects accordingly, we will understand by a side condition an *O*-objective for the class of decision situations generated in this way, whose reduced set of consequences consists of exactly two elements. Let $\mathfrak{A}'$ be the mentioned class; every side condition $N_i = (H_i, Q_i, g_i)$ $(i \varepsilon I)$ generates an objective function $z_i : \mathfrak{A}'^* \to \{0,1\}$. The set $\bar{\mathfrak{A}}_i^* := \{(A,a) \varepsilon \mathfrak{A}'^* \mid z_i(A,a) = 1\}$ has to be identified with the set of all actions feasible with respect to side condition N_i. H_i and Q_i incorporate the information derived from a) through d) and convert it to a (0,1)-statement.

As we have seen, side conditions of type B are something separate. They are to be generated by the following process: given an objective (H,Q,g) whose related reduced set of consequences usually consists of more than two elements, the relation Q is substituted by $Q^{\cdot} \subset Q$ and g by $g^{\cdot}$ such that $(H, Q^{\cdot}, g^{\cdot})$ becomes an objective, whose reduced set of consequences consists of exactly two elements. Hence, certain informations though provided by H are not exploited by $Q^{\cdot}$. Accordingly, to construct side conditions of type B means to waste information.

Constructing side conditions of type B can be substantiated by two arguments. The first is based on the assertion of the existence of *levels of aspiration.*[1)2)] The theory of the levels of aspiration asserts the behavior of real subjects to be indescribable by the classical assumption of a maximizing behavior with respect to a reduced set of consequences with more than two elements, where all action opportunities have

1)Simon [1957, p. 246]
2)Sauermann and Selten [1962, p. 577]

at the same time been placed at the decision maker's disposal: but rather by a successive search for a "satisfying" alternative working up the decision situation and decision making amalgamate to one and the same process.[1] This assertion might prove right in situations which are characterized by an extraordinarily high degree of complexity compared with the computational and other abilities of information attainment and exploitation of information. But the level of aspiration is suspected not to be absolutely fixed but rather to depend on the special decision situation (i.e. especially on the provided information);[2] therefore, if anything, those levels are to be interpreted as thresholds passing over which the individual does not believe a further information attainment to be profitable; this belief may be understood as a consequence of a pre-rational evaluation of the expected increase of the objective-achievement in the given situation. Hence, by the assumption made above - the assumption of a given information level and a given set of considered actions - the existence of a level of aspiration in Simon's terms cannot be ascertained.

A second kind of argumentation for the existence of conditions of type B is based on the assertion that they result from a compromise between conflicting objectives.[3] In fact that may be looked upon that way (under special assumptions as will be seen later), but it must be remarked on that a very special kind of priority and relative weight of the objectives is presupposed. But this evaluation of the relationships between the different objectives is a process which needs to be revealed and to be investigated; an embracing decision model should not allow most important decisions to have the character of data at all.

The preceding arguments set us to relieve side conditions of type B of their function as side conditions, i.e. we do not allow them to test the feasibilty of action, but rather consider the underlying objectives explicitely to be objectives. For investigating the decisions under uncertainty this standpoint will be of special importance.

---

1) Simon [1957, pp. 250-252]
2) Accordingly, Sauermann and Selten [1962] try to develop a theory of adjusting the levels of aspiration to new decision situations and new information respectively.
3) An approach to derive levels of aspiration within the framework of the chance-constrained programming may be found in Näslund [1967, pp. 17-20]; opportunities of substituting objective-achievements by security aspirations are considered in Albach and Schüler [1970].

1.3. *The Concept of Objectives under Uncertainty*

The previous assumption making a decision situation under uncertainty is described by $\Gamma=(A,\langle\Omega,\mathcal{A},\mu\rangle_A)$. In the following the set of final states will be denoted by $\Phi(\Gamma):=\bigcup_{a\varepsilon A}\{a\}\times\Omega_a$. The phenomenon of uncertainty does not only consist of the decision maker's ignorance, which final state will be realized when a certain action is selected, but additionally of the eventuality that the property described by a certain aspect might not be known for all possible final states: the relation Q of an aspect (n,Q) on $\Phi(\Gamma)$ is only known on a certain subset E(n) of $\Phi(\Gamma)$, i.e. the decision maker is informed on $Q\cap E(n)\times E(n)$ only. Final states in $\Phi(\Gamma)$ have a dimension with respect to the aspect (n,Q), i.e. the decision maker is informed on final states from $\Phi(\Gamma)$ with respect to (n,Q), if and only if they are contained in E(n). Hence by the decision maker's information level a subset $E(n)\subset\Phi(\Gamma)$ and an equivalence relation G(n) on E(n) are assigned to each name n of an aspect. We consider, as in the case of certainty, a class $\mathfrak{A}$ of decision problems of the general form Γ; analogously the coproduct $\mathfrak{A}^*:=\coprod_{\Gamma\varepsilon\mathfrak{A}}\Phi(\Gamma)$ is constructed to distinguish final states belonging to different decision situations. By Γ^* we denote the image of $\Phi(\Gamma)$ under inj_Γ, the related images of E(n) and G(n) will be denoted by the same symbols. With these constructions and notations in mind we are able to formulate a notion of a point of view under uncertainty analogously to the considerations under certainty. As in the case of certainty the point of view under uncertainty is an instrument of condensating and processing given information. Not all basic data which are on principle available to describe final states are relevant for a certain point of view; hence it is of no consequence whether or not there is information at hand on an irrelevant aspect; the absence of information is felt to be deficient only if it is needed. Accordingly, a point of view has to ascertain which aspects are relevant under the point of view in question: a point of view determines a mapping $T:\mathfrak{A}\to\mathfrak{P}(N)$ where N is the set of all names of aspects arising in $\mathfrak{A}$; T assigns to each situation the associated set (of names) of relevant aspects. Being an instrument of information processing the point of view, but only that information at hand, Γ^* has to be replaced by the set $E_{T(\Gamma)}:=\bigcap_{n\varepsilon T(\Gamma)}E(n)$. Only final states the decision maker is informed on under all relevant aspects can be investigated under the mentioned point of view. The elements of $E_{T(\Gamma)}$ again may be represented as multi-dimensional objects (for each relevant aspect one dimension): the representation-mapping

$$s_{T(\Gamma)} : E_{T(\Gamma)} \to \prod_{n\varepsilon T(\Gamma)} E(n)/G(n)$$

is defined such that the following diagram becomes commutative for all $m\varepsilon T(\Gamma)$:

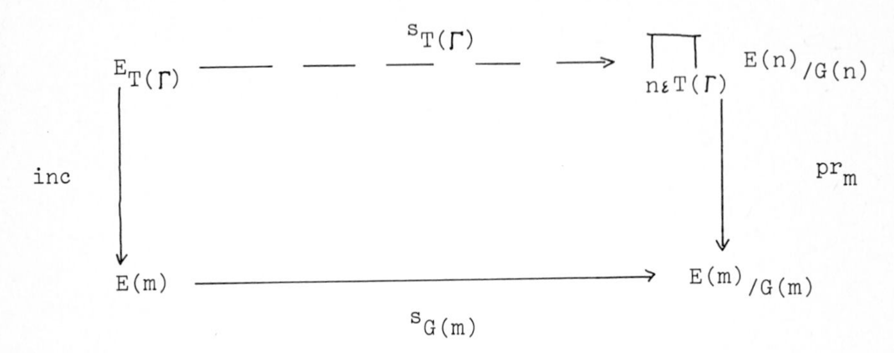

As in case of certainty the *data-dimension space* $\bar{\mathfrak{a}}_T = \bigcup_{\Gamma\varepsilon\mathfrak{a}} \prod_{n\varepsilon T(\Gamma)} E(n)/G(n)$ and a natural representation mapping $\Delta_T : \bigcup_{\Gamma\varepsilon\mathfrak{a}} E_{T(\Gamma)} \to \bar{\mathfrak{a}}_T$ are defined: $\Delta_T(\Gamma,a,\omega) := s_{T(\Gamma)}(a,\omega)$; under uncertainty both expressions depend on the considered points of view. Analogously to def. 1.2. we give the following

Definition 1.5. A *point of view under uncertainty* for the class $\mathfrak{a}$ is a pair (T,γ) consisting of a mapping $T : \mathfrak{a} \to \mathfrak{P}(N)$ and an element $\gamma\varepsilon PV(\mathfrak{a},T) := \coprod_{k\varepsilon N} X_k$ where $X_k = [\bar{\mathfrak{a}}_T]$ for all $k\varepsilon N$. We introduce the following notation:

$$\mathfrak{a}(T) := \bigcup_{\Gamma\varepsilon\mathfrak{a}} E_{T(\Gamma)} \qquad \text{and } \mathfrak{a}_{T_\gamma} := \bar{\mathfrak{a}}_{T/\gamma}.$$

The interpretation is analogous to the case of certainty: virtually, A has to be replaced by $\bigcup_{a\varepsilon A} \{a\}\times\Omega_a$; complications only arise because, for reasons of poor information, sometimes only a proper subset of final states can be considered under the point of view in question.

We are now prepared to carry over the concept of an objective under certainty to the described situation of uncertainty leading ourselves to the concept of a criterion or partial objective:

Definition 1.6. Given a class $\mathfrak{a}$ of decision situations under uncertainty a *V-criterion (or partial V-objective)* for $\mathfrak{a}$ is the equivalence class of a tripel (H,Q,g) meeting the following conditions:

(i) V is a subcategory of O.
(ii) H is a finite set of points of view under uncertainty for $\mathfrak{A}$.
(iii) Q is a complete preference relation on $S(H) := \prod_{(T,\gamma)\varepsilon H} \mathfrak{A}_{T_\gamma}$.
(iv) $g: S(H)_{/Q'} \to R$ is a mapping such that $(S(H)_{/Q'}, g)$ becomes an object of the category V.
(v) Let $\Gamma = (A, <\Omega, A, \mu>_A) \varepsilon \mathfrak{A}$ be given; for each $a \varepsilon A$

$$X_{\Gamma,a} := \{\omega \varepsilon \Omega_a | (\Gamma,a,\omega) \varepsilon \bigcap_{(T,\gamma)\varepsilon H} E_{T(\Gamma)}\} \ \varepsilon \ A_a$$

holds.

Two tripels meeting (i) through (v) are called *equivalent* if the following conditions are held:

(vi) $\bigcap_{(T,\gamma)\varepsilon H_1} \mathfrak{A}(T) = \bigcap_{(T,\gamma)\varepsilon H_2} \mathfrak{A}(T)$
(vii) Condition (v) of def. 1.4. holds formally, if it is replaced by $\bigcap_{(T,\gamma)\varepsilon H} \mathfrak{A}(T)$ and E_{HQ} by $S(H)_{/Q'}$.

Analogously to definition 1.4. we define S(H) to be the *set of consequences* and ψ_H to be the *result function:*

$$\bigcap_{(T,\gamma)\varepsilon H} \mathfrak{A}(T) \xrightarrow{\psi_H} S(H) \xrightarrow{\pi} S(H)_{/Q'}.$$ [1)]

To each situation $\Gamma \varepsilon \mathfrak{A}$ let be defined a formal element O_Γ; by the *criterion function* we mean the function defined as follows:

$$Z : \mathfrak{A}^* \to R \ \{O_\Gamma | \Gamma \varepsilon \mathfrak{A}\}$$

$$Z(\Gamma,a,\omega) = g \cdot \pi \cdot \psi_H(\Gamma,a,\omega) \text{ for final states}$$

$$(\Gamma,a,\omega) \ \varepsilon \bigcap_{(T,\gamma)\varepsilon H} E_{T(\Gamma)}$$

$$Z(\Gamma,a,\omega) = O_\Gamma \text{ elsewhere.}$$

1) A complete construction of the result function will be given within the proof of proposition 2.1.

By the rule

$Z' = 0$, if a final state is not contained in $\bigcap_{(T,\gamma)\in H} \mathfrak{A}(T)$, i.e. if it has no dimension with respect to some relevant aspects,

$Z' = 1$ elsewhere

another criterion is defined, the so-called *dual criterion*. Its criterion function describes whether final states occur (do not occur) which can (cannot) be evaluated by the criterion (H,Q,g).

In consequence of condition (v) it may be spoken of the probability of the occurence of final states which can be evaluated; this probability equals the expected value of the dual criterion Z'.

The concept of a criterion or partial objective generalizes the concept of an objective under certainty to the case of uncertainty by numerically evaluating the final states of a situation under uncertainty in a way equivalent to the case of certainty; the method is somewhat different, because certain final states may under uncertainty be excluded from evaluation for reasons of lacking information; on the other hand, and that is the reason we speak of a criterion and do not simply speak of an objective, the evaluation of final states under uncertainty is not an evaluation of *actions* as in case of certainty but rather an evaluation of states. Based on the evaluation of final *states* an evaluation of *actions* will become possible by a further step first: virtually, this is the problem of classical decision theory.[1] With regard to this problem we will take the standpoint most frequently taken by the literature: the standpoint of a rational behaviour in the sense of the Bernoulli -principle (expected-utility principle).[2]

Definition 1.7. A *Bernoulli-objective* for the class $\mathfrak{A}$ of decision situations under uncertainty is a *V*-criterion for $\mathfrak{A}$: (H,Q,g) with the following properties:

(i) *V* is a D-category where D is a certain subgroup of $L_+(R)$.

(ii) g is a bounded mapping.

---

1) For references we refer to Schneeweiß [1967].

2) A detailed presentation of this principle will be given in the following excursus.

(iii) The mapping $\pi\cdot\psi_H(\Gamma,a,.)$ is measurable with respect to the natural interval-σ-algebra of the reduced set of consequences.[1]

For each action a of the situation $\Gamma=(A,<\Omega,\mathcal{A},\mu>_A)$ meeting the condition $\mu_a(X_{\Gamma,a}) \neq 0$ the following integral is defined:

$$\int_{X_{\Gamma,a}} g\cdot\pi\cdot\psi_H(\Gamma,a,.)\; d(\mu_a|X_{\Gamma,a}).^{2)}$$

$\mu_a|X_{\Gamma,a}$ is the conditional probability measure on $X_{\Gamma,a}$: we will only work on the subspace of final states having a dimension with respect to *all* relevant aspects. The function defined by the integral on the set of considered actions will be called the *Bernoulli-objective function*. This function assigns to each action the conditional utility-expectation related to this action within a certain decision situation.

1.3.1. Excursus: Axiomatical Treatment of the Bernoulli-Principle Some Results on Continuity and Integrability of Utility Functions.[3]

This excursus is to develop the Bernoulli-principle and to connect it with the chosen concept of a Bernoulli-objective. Here the following problem arises: the concept of a Bernoulli-objective presupposes the structure of a preference ordering on the set of consequences; the Bernoulli-principle, however, consists in operationally representing a preference ordering on the set of probability distributions of consequences; it makes use of the expected value of a utility function on the set of consequences. Hence we have to make clear the conditions, under which the expected value of utility on an abstract set of consequences may be formed, i.e. conditions of integrability.

On the other hand a natural requirement of utility functions demands that "small" changes of the consequences do not generate any "great" changes of evaluation at random, a requirement the concept of a Bernoulli-

1) For the concept of the interval-σ-algebra we refer to the following excursus.
2) A proof of the integrability of g will be given in corollary of lemma 1.3..
3) The first axiomatical treatment of the Bernoulli-principle is to be found in the fundamental work of von Neumann and Morgenstern [1946].

objective fulfils just by construction; accordingly, conditions of continuity have to be clarified. Since both problems, as far as we can see, have not yet been treated in a manner suitable to our questioning, namely with respect to abstract (not measured by numerical terms) sets of consequences, we are induced to this excursus.

For a treatment of the mentioned problems we will proceed in three steps: the first will deal with questions of the measurable structures generated by the ordering structure; the second with the generated topological structures;and the third with the connection between both structures regarding the intended application to a representation of a preference ordering on the set of probability distributions of consequences by an integral of a continuous utility function on the set of consequences. The continuity of the utility function will be ensured by a new assumption, for in the literature this aspect is not sufficiently taken into account. The results obtained are partly new - at least in presented form.

1.3.1.1. The Natural σ-Algebra and the Interval-Topology

1.3.1.1.1. Measurable Structures

Let E be an arbitrary set provided with a preference ordering $\preccurlyeq$ whose associated indifference relation is denoted by $\sim$; the relation $\preccurlyeq\backslash\sim$ ("strictly prefered") will be denoted by $\prec$. The quotient set $E_{/\sim}$ is totally ordered in a natural way and will be denoted by $\bar{E}$, the associated ordering by $\leq$. Especially, $\leq$ is a preference relation too; hence definitions for E and $\preccurlyeq$ hold analogously for $\bar{E}$ and $\leq$; these facts will be used without further remarks.

We define *intervals* in E: let a and b be some elements of E.

(1) $<a,b>:=\{\alpha\varepsilon E \mid a \preccurlyeq \alpha \preccurlyeq b\}$ (2) $<a,b<:=\{\alpha\varepsilon E \mid a \preccurlyeq \alpha \prec b\}$

(3) $>a,b>:=\{\alpha\varepsilon E \mid a \prec \alpha \preccurlyeq b\}$ (4) $>a,b<:=\{\alpha\varepsilon E \mid a \prec \alpha \prec b\}$

(5) $<a,\rightarrow<:=\{\alpha\varepsilon E \mid a \preccurlyeq \alpha\}$ (6) $>a,\rightarrow<:=\{\alpha\varepsilon E \mid a \prec \alpha\}$

(7) $>\leftarrow,a>:=\{\alpha\varepsilon E \mid \alpha \preccurlyeq a\}= >a,\rightarrow<^c$ (8) $>\leftarrow,a<:=<a,\rightarrow<^c$

(9) $>\leftarrow,\rightarrow<:=E$

The superscript ^{c} indicates the formation of the complement in E.

The intervals of the types (1) through (4) are called *bounded*, the remaining ones *unbounded* (Remark: if there exist greatest or smallest elements in E, there are intervals at the same time bounded and unbounded!). We mention some intervals of special nature: $>a,a< \, = \emptyset$ for all $a \in E$; the intervals $<a,a>$ are exactly the elements of $\bar{E}$. Denoting by $\mathcal{S}(E)$ the set of all intervals on E there is a bijective mapping $\psi : \mathcal{S}(E) \rightarrow \mathcal{S}(\bar{E})$ defined by $\psi(S)=\{<a,a> \in \bar{E} \mid a \in S\}$ for all $S \in \mathcal{S}(E)$; this is immediately seen. Clearly the set $\mathcal{S}(E)$ is a Boolean semi-algebra.[1)]

As is seen from def. 1.6. we considered the totally ordered reduced set of consequences instead of the preference ordered set of consequences; the reason is that totally ordered sets are mathematically more easily dealt with. But not to allow this approach to have substantial consequences it has to be shown that certain relevant structures are isomorphically transported to the reduced set of consequences.

Lemma 1.1. Let Ω be some set and $\mathcal{A}$ a Boolean algebra on Ω. Given an atom $A \in \mathcal{A}$ (i.e. $\emptyset \neq B$ and $B \subset A$ and $B \in \mathcal{A}$ together imply $B = A$), A is an atom for the σ-algebra $\mathcal{A}'$ generated by $\mathcal{A}$.

Proof: Let $A \in \mathcal{A}$ be an atom of $\mathcal{A}$ and $\mathcal{B}:=\{B \in \mathcal{A}' \mid \emptyset \neq B \; A \neq A\}$. Supposing A not to be an atom in $\mathcal{A}'$ the class $\mathcal{B}$ is non-empty. Forming $\mathcal{C}:=\mathcal{A}' \setminus \mathcal{B}$ and showing that $\mathcal{C}$ is a σ-algebra we have $\mathcal{C} = \mathcal{A}'$, because $\mathcal{A} \subset \mathcal{C} \subset \mathcal{A}'$ holds and $\mathcal{A}'$ is generated by $\mathcal{A}$; hence $\mathcal{B}$ is empty and A is not an atom of $\mathcal{A}$ contrary to the assumption. Hence A is an atom in $\mathcal{A}'$. It remains to be shown that $\mathcal{C}$ is a σ-algebra. The proof is given in two steps:
Assertion 1: $\mathcal{C}$ is an algebra.
a) $\emptyset \in \mathcal{C}$ and $\Omega \in \mathcal{C}$, that is obvious.
b) Let be $C \in \mathcal{C}$; then $C \cap A$ is empty or equals A. It is easily seen that $C^c \cap A$ is either empty or equals A from what we have $C^c \in \mathcal{C}$.
c) Using the defining condition of $\mathcal{C}$ it is easily established that the union of two elements of $\mathcal{C}$ belongs to $\mathcal{C}$ too.
a) through c) together prove assertion 1.
Assertion 2: $\mathcal{C}$ is a monotone class.[2)]
d) Let C^i be a monotone increasing sequence of elements of $\mathcal{C}$ $(i=1,2,..)$. If there is an index j with $C^j = A$ we have $\bigcup_{i \in N} (C^i \cap A) = (\bigcup_{i \in N} C^i) \cap A = A$; if for all $i \in N$ we have $C^i \cap A = \emptyset$ the intersection of A and the union of all C^i must be empty too from what assertion 2 follows, if the dual case of monotone decreasing sequences is proved analogously. A detailed proof may be left to the reader. ¶

By $\mathcal{E}$ we denote the σ-algebra generated by $\mathcal{S}(E)$ and by $\bar{\mathcal{E}}$ the σ-algebra generated by $\mathcal{S}(\bar{E})$.

---

1) For a definition we refer to Neveu [1969, p. 41], Definition 1.6.1.
2) Neveu [1969, p. 27], Definition 1.4.2.

Proposition 1.2. $\mathcal{E}$ and $\bar{\mathcal{E}}$ are isomorphic.

Proof: E is completely decomposed by the sets $\langle a,a\rangle$ where $a\varepsilon E$; these sets are exactly the atoms of the Boolean algebra generated by $\mathcal{S}(E)$[1] which are by lemma 1.1. the atoms of $\mathcal{E}$. For all $A\varepsilon\mathcal{E}$ and all $a\varepsilon E$ the intersection $A\cap\langle a,a\rangle$ is either empty or equals $\langle a,a\rangle$. The mapping ψ defined above may be extended to all of $\mathcal{E}$ formally using the same definition, it becomes, then, a mapping from $\mathcal{E}$ into $\mathfrak{P}(\bar{E})$. In the converse direction (from $\mathfrak{P}(\bar{E})$ into $\mathfrak{P}(E)$) the mapping ψ' may be defined using the formula $\psi'(\bar{A}) = \bigcup_{\langle a,a\rangle\varepsilon\bar{A}} \langle a,a\rangle$. The following equations hold: $\psi'\cdot\psi = \text{id}$ and $\psi\cdot\psi'|_{\text{Im }\psi} = \text{id}|_{\text{Im }\psi}$.

Assertion: Im ψ is a σ-algebra and ψ is a homomorphism.
This is easily seen by using the definition; in the same way is $\psi'|_{\text{Im }\psi}$ seen to be a homomorphism too; accordingly both mappings are isomorphisms. It remains to be shown that Im $\psi = \bar{\mathcal{E}}$ holds. Since ψ transforms the intervals of $\mathcal{S}(E)$ into the intervals of $\mathcal{S}(\bar{E})$, $\bar{\mathcal{E}}\subset\text{Im }\psi$ must hold, because $\bar{\mathcal{E}}$ is generated by $\mathcal{S}(\bar{E})$. On the other hand ψ' exactly transforms the intervals of $\mathcal{S}(\bar{E})$ into the intervals of $\mathcal{S}(E)$; accordingly $\mathcal{E}\subset\psi'(\bar{\mathcal{E}})$ and $\psi(\mathcal{E})\subset\psi\cdot\psi'(\bar{\mathcal{E}}) = \bar{\mathcal{E}}$ hold together implying what we want. This proves the proposition. ¶

If we want to consider probability measures on reduced sets of consequences instead of those on sets of consequences, we have to go on a step further:

Corollary: Let $\mathcal{M}(E,\mathcal{E})$ be the set of probability measures on the measurable space $(E,\mathcal{E})$ and $\mathcal{M}(\bar{E},\bar{\mathcal{E}})$ the corresponding set on $(\bar{E},\bar{\mathcal{E}})$. For each $\mu\varepsilon\mathcal{M}(E,\mathcal{E})$ we define

$$\alpha(\mu)(\psi(S)) := \mu(S) \qquad \text{for all } S\varepsilon\mathcal{S}(E).$$

Then α may be extended to a well-defined bijective mapping from $\mathcal{M}(E,\mathcal{E})$ onto $\mathcal{M}(\bar{E},\bar{\mathcal{E}})$ meeting the condition

$$\alpha(\mu)(\psi(A)) = \mu(A) \qquad \text{for all } A\varepsilon\mathcal{E} \text{ and}$$

all $\mu\varepsilon\mathcal{M}(E,\mathcal{E})$.

Proof: This is an immediate consequence of prop. 1.2. and Satz 1.6.1. of Neveu.[2] ¶

By the corollary of prop. 1.2. it is ensured that statements on proba-

1) Neveu [1969, p. 41], Satz 1.6.1.
2) Neveu [1969, p. 41].

bility measures on the reduced set of consequences of a Bernoulli-objective completely correspond to associated statements on those which are on sets of consequences.

1.3.1.1.2. Topological Structures

In this paragraph we will analyse the topological structure generated by a preference relation.

Definition 1.8. Intervals of the types (4), (6), (8) and (9) are called *open*, those of the types (1), (5), (7) and (9) are called *closed*.

A subset S of E is called a *segment* in E if for all $a,b \in S$ the interval $\langle a,b \rangle$ is contained in S. Especially all intervals in E are segments in E.

Given an element x of E the set of open intervals containing x forms a filter-base of neighbourhoods of x; the family of all such filter-bases defines a unique topology for E, the so-called *interval topology* denoted by $\mathcal{J}_E$.[1] Naturally the interval topology of $\bar{E}$ is identical with the quotient topology induced by E and the indifference relation.

In the following the connection between the relative topology and the interval topology on a subspace of a preference ordered set is dealt with. To this intention we throughout presuppose X to be some set provided with a preference relation $\preccurlyeq$ and P to denote a subset of X, also preference ordered by the restriction of $\preccurlyeq$ to P.

Definition 1.9. Let P_1 and P_2 be nonempty subsets of X mutual complements to each other in X, additionally meeting the following requirement:

$$x \in P_1 \text{ and } y \in X \text{ and } y \preccurlyeq x \text{ imply } y \in P_2.$$

The pair (P_1, P_2) is called a *Dedekind-decomposition* of X.[2] A Dedekind-decomposition is a decomposition of X in two nonempty segments.

---

1) Bourbaki [1965, chap. 1, § 1, No 2].
2) Lenz [1961, p. 44].

By a *gap* is understood a Dedekind-decomposition (P_1,P_2) where neither of the segments is a closed interval. By a *step* we mean a Dedekind-decomposition such that both segments are closed intervals. Finally, by a *cut* is meant a Dedekind-decomposition (P_1,P_2) where either P_1 or P_2 is a closed interval.[1]

Definition 1.10. P will be called *simple in* X[2] if every Dedekind-cut (P_1,P_2) of P generates a Dedekind-decomposition of X by virtue of

$$P_1' := \{y\varepsilon X \mid y \preccurlyeq x \text{ for at least one } x\varepsilon P_1\}$$

$$P_2' := \{x\varepsilon X \mid y \preccurlyeq x \text{ for at least one } y\varepsilon P_2\}.$$

The generated decomposition is a cut of X, for, given P_1 to be a closed interval in P the set P_1' is a closed interval in X; but P_2' cannot be a closed interval in X because $P_2' = \langle a,\rightarrow\langle$ would imply that there exists a $y\varepsilon P_2$ with $y \lessdot a$ and hence $y \sim a$ and thus $P_2 = \langle y,\rightarrow\langle$ in P contrary to the assumption. The other case is analogous.

Proposition 1.3. (due to Pfanzagl)[3] Given X and P as above, the interval topology of P is identical with the relative topology of P in X (X provided with its interval topology) if, and only if, P is simple in X.

To prove the proposition we first give a lemma:

Lemma 1.2. The relative topology of P in X is finer than the interval topology $\mathcal{J}_P$ of P.

Proof: It is clear that for every open interval in P there is the trace of an open interval in X on P. ¶

Proof of Proposition 1.3.: 1) Let P be simple in X; it has to be shown that the interval topology $\mathcal{J}_P$ is finer than $\mathcal{J}_X \cap P$. Let I be an open interval in X and $I \cap P \neq \emptyset$. It has to be shown that there is an open interval L in P with $L \subset I \cap P$, for the set of open intervals forms a base for the interval topology. Denoting $Q := I \cap P$ we choose an element $x\varepsilon Q$ and construct the following two decompositions of P:

$$P_1 = \{y\varepsilon P \mid y \preccurlyeq x\}, \quad P_2 = P\setminus P_1 \quad \text{and} \quad T_2 = \{y\varepsilon P \mid x \preccurlyeq y\}, \quad T_1 = P\setminus T_2.$$

---

1) Lenz [1961, p. 44]; gap = "Lücke", step = "Sprung", cut = "Schnitt".
2) Pfanzagl [1968, p. 62].
3) Pfanzagl [1968, p. 62], 3.4.5. Lemma.

At first we suppose P_2 and T_1 to be non-empty, then (P_1,P_2) and (T_1,T_2) are cuts or steps but no gaps.

a) If (P_1,P_2) is a cut we have a $z \varepsilon Q \cap P_2$ since P is simple in X; from $I_2 := <x,z<_P$ we have $I_2 \subset Q$.

b) If (T_1,T_2) is a cut there is a $y \varepsilon Q \cap T_1$: $I_1 := >y,x>_P$.

c) If (P_1,P_2) is a step there must be a smallest element $z \varepsilon P_2$ and we have $I_2 := <x,z<_P \subset Q$.

d) If (T_1,T_2) is a step there must be a greatest element $y \varepsilon T_1$ and we have $I_1 := >y,x>_P \subset Q$.

In all instances the following holds: $L := I_1 \cap I_2 = >y,z<_P \subset Q$ is an open interval of P containing x and completely contained in Q. If T_1 or P_2 are empty y (z respectively) has to be substituted by $\leftarrow$ ($\rightarrow$ respectively.

2) Let $\mathcal{J}_P = \mathcal{J}_X \cap P$ and (P_1,P_2) be a cut in P. Without any loss of generality let P_1 be the closed interval. Then, following def. 1.10. P_1' is a closed interval and P_2' an open segment in X. Let $x \varepsilon X \setminus (P_1' \cup P_2')$, then $x \notin P$ and $P_1 \prec x \prec P_2$ hold. The set $P_x := \{y \varepsilon X \mid y \prec x\}$ is open in X; hence $P_1 = P_x \cap P$ is relatively open and consequently open with respect to the interval topology of P. Thus, there is an open interval of P containing the greatest element of P_1 and contained in P_1. This can happen only if there is a smallest element in P_2. But then (P_1,P_2) is no longer a cut in P. Accordingly, we have $P_1' \cup P_2' = X$ and thus (P_1',P_2') is a cut in X. ¶

Proposition 1.4. Let X be totally ordered. The interval topology $\mathcal{J}_X$ of X is discrete, if, and only if, X has no Dedekind-cuts.

Proof: 1) Let the interval topology of X be discrete.

Let (P_1,P_2) be a Dedekind-decomposition of X. Without any loss of generality we may assume P_1 to be the closed interval in X. As the topology of X is discrete, the set P_1 is open. Let $P_1 := >\leftarrow,x>$. Then there is an open interval $I = >a,b<$ with $I \subset P_1$ and $x \varepsilon I$. Consequently we have $b \varepsilon P_1$ and thus $P_2 = <b,\rightarrow<$ and hence (P_1,P_2) cannot be a cut.

2) Suppose X to have no Dedekind-cuts.

Let I be an open interval of the form $>a,\rightarrow<$. If there is no smallest element in I, $(X \setminus I, I)$ is a cut contrary to the assumption. Hence there is a smallest element in I and I is closed. Analogously one has to argue with respect to intervals of the form $>\leftarrow,a<$. Hence all open unbounded intervals are closed sets. Accordingly, all sets consisting of one element are at the same time open and closed which proves the assertion.

By 1) and 2) the proposition is completely proved. ¶

Proposition 1.5. Let X be preference ordered and P be a subset of X. If P is connected with respect to its relative topology, P is simple in X.

Proof: Let (P_1,P_2) be a Dedekind-cut in P. P_1' and P_2' are defined as in def. 1.10.. If (P_1',P_2') is no decomposition of X, there is a $x\varepsilon X$ with $P_1 \prec x \prec P_2$ and consequently $P_x := \{y\varepsilon P | y \prec x\} = P_1$ and $P^x := \{y \varepsilon P | x \prec y\} = P_2$ hold in this case, and (P_1,P_2) is a relatively open decomposition of P contrary to the assumption that P is connected in X. Hence (P_1',P_2') is a cut in X and P is simple in X. ¶

Proposition 1.6. Every segment in X is simple in X.

The proof is obvious.

Definition 1.11. Let X and Y be preference ordered sets. A mapping f from X to Y is called an order homomorphism if $a \preccurlyeq b$ implies $f(a) \preccurlyeq f(b)$.

An order homomorphism f is called *order-faithful* if

$$f(a) \sim f(b) \Rightarrow a \sim b.$$

Proposition 1.7. Let X and Y be preference ordered sets and f be an order-faithful homomorphism from X to Y. f is continuous with respect to the interval topology of X and Y if, and only if, Im f is simple in Y.

Proof: 1) Let f be continuous and (P_1,P_2) be a cut of Im f. Suppose that an element $y\varepsilon Y$ exists with $P_1 \prec y \prec P_2$; then the sets P_1 and P_2 are relatively open. Because of the continuity of f the sets $f^{-1}(P_1)$ and $f^{-1}(P_2)$ are open in X. As f is order-faithful these sets form a cut in X. We may suppose that $f^{-1}(P_1)$ is the closed interval. This interval is an open set if, and only if, there exists a smallest element in $f^{-1}(P_2)$. Contrary to the assumption $(f^{-1}(P_1),f^{-1}(P_2))$, then, cannot be a cut. This proves that Im f is simple.

2) Let Im f be simple in Y. Then the relative topology and the interval topology of Im f coincide. Let A be an open set in Y. $A \cap \text{Im } f$ is at the same time a relatively open and an interval-open set in Im f, then there is an interval I open in Im f and contained in A. As f is order-faithful, $f^{-1}(I)$ is an open interval in X. This immediately implies the continuity of f. ¶

1.3.1.1.3. Connections between the Measurable and the Topological Structures of Preference Ordered Sets

Theorem 1.1. Let X be a preference ordered set and T a topology on X with the following properties:

(i) T is finer than the interval topology J_X.
(ii) There is a countable base of T.

Then the following holds:

The σ-algebra E on X generated by the semi-algebra $S(X)$ of intervals in X coincides with the Borel-σ-algebra of the interval toplogy of X denoted by B.

Proof: Obviously we have $E \subset B$ because of $S(X) \subset B$. It remains to be shown $B \subset E$. Let A be a J_X-open set. There is a countable base of A as a T-subspace. The set of all intervals contained in A is a J_X-open covering of A. As T is finer than J_X this covering is T-open too. By Prop. 13 of Bourbaki [1958, chap. 9, § 2, No 8] there is a countable subcovering $\{I_n\}_{n \in N}$. Hence $A = \bigcup_{n \in N} I_n$ holds. For $I_n \in E$ for all $n \in N$, $A \in E$ holds. Thus E contains all J_X-open sets. This completes the proof. ¶

Corollary 1 Let X meet the conditions of theorem 1.1.. If an order-faithful homomorphism f from X into the real numbers is continuous, it is measurable (continuity with respect to the interval topology, measurability with respect to the interval-σ-algebra). If, additionally, f is bounded, it is integrable.

Proof: An order-faithful homomorphism f is as a mapping $f : X \rightarrow \text{Im } f$ an open mapping. Since R and thus the subspace Im f have countable bases, there is a countable base of X too in consequence of the continuity of f. By theorem 1.1. E coincides with the Borel-σ-algebra. By prop. 10 of Bourbaki [1958, chap. 9, § 6, No 3] f is measurable with respect to E and, if bounded, integrable. ¶

Corollary 2 Let X be preference ordered and J_X have a countable base. And let $P \subset X$ and F be the interval-σ-algebra of P. Then the Borel-σ-algebra of the interval topology of P coincides with F.

The proof may be left to the reader.

Lemma 1.3. Let X be preference ordered. Every segment in X belongs to the Borel-σ-algebra of the interval topology of X.

Proof: Let S be a segment in X. We define

$P_1 := \{x \varepsilon X | x \preccurlyeq y \text{ for at least one } y \varepsilon S\}$ $\qquad P_2 := (P_1)^c$

$T_2 := \{x \varepsilon X | y \preccurlyeq x \text{ for at least one } y \varepsilon S\}$ $\qquad T_1 := (T_2)^c$

Obviously (P_1, P_2) and (T_1, T_2) are Dedekind-decompositions of X and we have $S = P_1 \cap T_2$. If one of the four sets P_1, P_2, T_1 or T_2 is a closed interval, this set is closed with respect to the interval topology of X; if one of the four sets, say P_1, is not a closed interval it must be open with respect to the interval topology of X, for $x \varepsilon P_1$ implies that there is an element $y \varepsilon S$ with $x \preccurlyeq y$; there even must be a $y \varepsilon S$ with $x \prec y$, for otherwise we would have $P_1 = >\leftarrow, x>$ contrary to the assumption. Hence P_1 is open and S is the intersection of two sets being open or closed and thus Borelian.

1.3.1.2. *The Expected-Utility-Theorem*

In this paragraph P is a totally ordered set provided with the order relation $\leq$. Let E be the interval-σ-algebra of P. By $X := M(P, E)$ we denote the set of all probability measures on the measurable space (P, E).

Lemma 1.4. Let $\preccurlyeq$ be a complete preference relation on X. Let $i : P \rightarrow X$ be an imbedding map defined as follows:

For all $x \varepsilon P$ and all $E \varepsilon E$: $\quad i_x = \begin{cases} 1 \text{ for } x \varepsilon E \\ 0 \text{ for } x \notin E \end{cases}$

where i_x is the image of x under i (i_x is a function defined on E). Let the following conditions be met:

(B1) i is an order-faithful homomorphism (P and Im i may be identified such that this condition can be reformulated: $\leq$ and the trace of $\preccurlyeq$ on P coincide).

(B2) $p, q \varepsilon X$ and $p \prec q$ imply that for all $\alpha \varepsilon >0,1<$ and each $r \varepsilon X$ we have $\alpha p + (1 - \alpha) r \prec \alpha q + (1 - \alpha) r$.

(B3) $p, q, r \varepsilon X$ and $p \prec q \prec r$ imply that there are real numbers $\alpha, \beta \varepsilon >0,1<$, such that $\alpha p + (1 - \alpha) r \prec q \prec \beta p + (1 - \beta) r$.

(B4) Suppose $C \varepsilon E$ and $p, q, r \varepsilon X$ and $p(C) = 1$. If for all $x \varepsilon C$ we have $q \prec i_x$ then $q \preccurlyeq p$ holds. If for all $x \varepsilon C$ we have $i_x \prec r$, then $p \preccurlyeq r$ holds.

Under these assumptions there is an order-faithful homomorphism $U : X \rightarrow R$ uniquely determined up to positive linear transformations, whose restriction $u := U|_{\text{Im } i}$ is bounded. Additionally U meets the condition (of *expected utility*):

$p,q \varepsilon X$ and $\alpha \varepsilon >0,1<$ imply $U(\alpha p + (1-\alpha)q) = \alpha U(p) + (1-\alpha)U(q)$.

For the proof we may refer to Fishburn.[1)]

Interpretation of the conditions

(B1): *Condition of monotony*
A safe prospect guaranteeing a result preferred to the result of another safe prospect will itself be preferred.

(B2): *Condition of monotonicity*
If a prospect q is preferred to a prospect p, any probability mixture of q and some prospect r will be preferred to the probability mixture of p and r; if the mixture is in both cases generated in the same way: in some sense the preference relation is compatible with the formation of probability mixtures.

(B3): *Condition of intermediate prospects*
A prospect q lying between two further prospects p and r (in the sense of preference ordering) and probability mixtures of p and r such that the respective mixtures will be less preferred to q.

(B4): *Condition of dominance*
A prospect q less preferred to each safe result of a set C of results will not be preferred less (nor respectively preferred) to a prospect, which can almost safely guarantee that the realized result is a member of C.

In the following P and Im i are identified.

Corollary U is bounded on X.

Proof: If there exists a greatest (smallest) element x in X, U(x) is defined and finite. Since U is order-faithful the map U is bounded from above (from below).

1) Fishburn [1970, p. 112 and p. 138], theorem 8.4 and lemma 10.5.

Suppose X to have no greatest (smallest) element. Assume an element $x \varepsilon X$ such that $y \prec x$ ($x \prec y$) holds for all $y \varepsilon P$. By the second part of (B4), replacing C by P one has $z \preccurlyeq x$ ($x \preccurlyeq z$) for all $z \varepsilon X$ which means that x is a greatest (smallest) element of X contrary to the assumption. Consequently there is at least one element $y \varepsilon P$ with $x \preccurlyeq y$ ($y \preccurlyeq x$) for all $x \varepsilon X$; hence $U(x) \leq u(y) < B$ ($B < u(y) \leq U(x)$) holds because of the boundedness of u. ¶

Lemma 1.5. Let X meet the conditions of lemma 1.4.. Then the following statements hold:

(1) $p,q \varepsilon X$ and $p \prec q$ and $0 \leq \alpha < \beta \leq 1$ imply $\beta p + (1 - \beta)q \prec \alpha p + (1 - \alpha)q$.

(2) $p,q,r \varepsilon X$ and $p \preccurlyeq q \preccurlyeq r$ and $p \prec r$ imply that there is one and only one $\alpha \varepsilon]0,1[$ such that $q \sim \alpha p + (1 - \alpha)r$.

For a proof we refer to Fishburn.[1]

Statement (2) of lemma 1.5. gives rise to the following definition: for $a,b \varepsilon X$ with $a \prec b$ we define a map $g_{ab} : X \rightarrow [0,1]$ by

$$g_{ab}(x) := \begin{cases} 1 & \text{if } x \preccurlyeq a \\ \alpha & \text{if } a \preccurlyeq x \preccurlyeq b, \ \alpha \text{ given by lemma 1.5.} \\ 0 & \text{if } x \succcurlyeq b \end{cases}$$

The family $\{g_{ab} | a \prec b\}$ defines a uniform structure on X, namely the weakest uniform structure making all maps g_{ab} uniformly continuous.[2] This uniform structure will be denoted by $\mathcal{U}$; it coincides with the uniform structure defined by the family $\{f_{ab} | a \prec b\}$ of quasi-distances where $f_{ab}(x,y) := |g_{ab}(x) - g_{ab}(y)|$. We will prove the following theorem:

Theorem 1.2. Let X meet the conditions of lemma 1.4.. The topology generated by $\mathcal{U}$ (this topology is denoted by $\mathcal{U}$ too) coincides with the interval topology $\mathcal{J}_X$ of X.

Proof: 1) $\mathcal{U}$ is finer than $\mathcal{J}_X$. We will show that for each $x \varepsilon X$ every $\mathcal{J}_X$-neighbourhood of x contains a $\mathcal{U}$-neighbourhood of x.

a) Let $x \varepsilon X$ be neither a smallest nor a greatest element of X. By definition every $\mathcal{J}_X$-neighbourhood of x contains an open interval with $x \varepsilon]a,b[$. Hence $0 < g_{ab}(x) < 1$. Let be $0 < \delta < \min\{\alpha, 1-\alpha\}$. Then $f_{ab}(x,y) < \delta$ implies $y \varepsilon]a,b[$, for: $y \preccurlyeq a \Rightarrow g_{ab}(y) = 1 \Rightarrow f_{ab}(x,y) = 1 - \alpha > \delta$ and $y \succcurlyeq b \Rightarrow g_{ab}(y) = 0 \Rightarrow f_{ab}(x,y) = \alpha > \delta$. The "sphere" with centre x and a radius less than δ is a $\mathcal{U}$-neighbourhood which meets our requirements.

---

1) Fishburn [1970, pp. 111-112], theorem 8.3, c 1 and c 2.
2) Bourbaki [1958, chap. 9, § 1, No 2]

b) Let $x \varepsilon X$ be a smallest element of X. Then every J_X- neighbourhood of x contains an open interval of the form $>\leftarrow, b<$ with $b \succ x$. Let be $\delta=1$ and $f_{xb}(x,y)=1-g_{xb}(y)<\delta$ for all $y \varepsilon >\leftarrow, b<$. Hence $>\leftarrow, b<$ itself is a U-neighbourhood of x. The proof in the case of x being a greatest element of X is analogous.

2) J_X is finer than U. Let x be an arbitrary element of X. A typical element V of a fundamental system of U-neighbourhoods of x may be described as follows:

$\delta>0$, $n \varepsilon N$; let $>a_i, b_i<$ be an open non-empty interval of type (4)($i=1,..,n$) then V may be defined as the set of all $y \varepsilon X$ with $f_{a_i b_i}(x,y)<\delta$ for all $i=1,\dots,n$. In the following we make use of the abbreviations $f_i=f_{a_i b_i}$ and $g_i=g_{a_i b_i}$.

Without any loss of generality we may choose $\delta<1$. Let i be some index $1 \leq i \leq n$.

a) $g_i(x)=0$: $y:=\delta a_i+(1-\delta)b_i$ and $I_i:=>y,\rightarrow<$

b) $g_i(x)=1$: $y:=(1-\delta)a_i+\delta b_i$ and $I_i:=>\leftarrow,y<$

c) $0<g_i(x)=\alpha<1$:

$$\chi := \begin{cases} \alpha - \delta & \text{if } \alpha - \delta > 0 \\ 0 & \text{elsewhere} \end{cases}$$

$$\psi := \begin{cases} \alpha + \delta & \text{if } \alpha + \delta < 1 \\ 0 & \text{elsewhere} \end{cases}$$

$$y_1:=\chi a_i+(1-\chi)b_i \qquad y_2:=\psi a_i+(1-\psi)b_i$$

$$I_i:=>y_1,y_2<$$

It is now easily seen that the intersection of all I_i is a J_X-neighbourhood of x which is contained in V. This proves assertion 2) and thus the theorem is seen to be completely proved. ¶

Corollary 1 The maps g_{ab} are continuous with respect to the interval topology of X.

This is obvious.

Corollary 2 The map U provided by lemma 1.4. is continuous with respect to the interval topology of X. Especially U is uniformly continuous on every bounded interval of X.

Proof: The set X is covered by the set of open and bounded intervals up to greatest or smallest intervals. The restriction of U to those intervals is seen to be uniformly continuous by using the property of expected utility and corollary 1. If there is a smallest element a or a greatest element b we choose y such that $a \prec y \prec b$ holds; the intervals $<a,y<$ and $>y,b>$ are open with respect to the interval topology of X. The restriction of U to these intervals is continuous too, X is completely covered and an application of Proposition 4 of Bourbaki [1965, chap. 1, § 3, No 2] completes the proof. ¶

Lemma 1.6. The image of a bounded open interval of X under U is a bounded open interval of R.

The proof is immediately seen by using the order-faithfulness, condition (B2) and lemma 1.5..

Proposition 1.11. X is a connected topological space.

Proof: It is easily seen that X is connected, if and only if every Dedekind-decomposition of X is a cut. Hence, let (P_1,P_2) be a Dedekind -decomposition of X; assume P_2 not to be a closed interval. Furthermore let be $a \varepsilon P_1$ and $b \varepsilon P_2$. By assumption we have $>a,b< \cap P_2 \neq \emptyset$. By lemma 1.4. and statement (2) of lemma 1.5. the image of $>a,b<$ is an open interval of R. If a is a greatest element of P_1, (P_1,P_2) is a cut in X. Thus, if it were not so, then $>a,b< \cap P_1 \neq \emptyset$ and $(U(>a,b< \cap P_1), U(>a,b< \cap P_2))$ is a Dedekind-decomposition of $U(>a,b<)$ since U is order-faithful. But in open intervals of R there are only cuts; hence by the order-faithfulness of U (P_1,P_2) is a cut too. This proves the proposition for X cannot have steps because of condition (B2) of lemma 1.4.. ¶

Corollary The image of X under U is a bounded interval of R.

This is clear.

Theorem 1.3. (Expected-utility-theorem; Bernoulli-principle) Let X meet the conditions of lemma 1.4.. Then u is integrable with respect to the interval-σ-algebra of P and we have for all $x \varepsilon X$:

$$U(x) = \int_P u \, dx.$$

Proof: The integrability of u is an immediate consequence of corollary 2 of theorem 1.2. and of corollary of lemma 1.3.. The remaining statements have been proved by Fishburn [1970, p. 142], 10.6..

Theorem 1.4. *(Continuity of the utility function)*
Let X meet the conditions of lemma 1.4.. Additionally we assume the following to hold:

(B5) Given a Dedekind-cut (P_1,P_2) of P there are for all $x \varepsilon P_1$ and $y \varepsilon P_2$ and $\alpha \varepsilon >0,1<$ elements $a,b \varepsilon P$ with

$$x \prec a \prec \alpha x + (1 - \alpha) y \prec b \prec y.$$

Then the interval topology of P coincides with the relative topology of P generated by X and u is a continuous function.

Proof: By (B5) P is simple in X. The rest follows from prop. 1.3 . ¶

We now add a further result:

Proposition 1.12. *(Existence of a certainty equivalent)*
X meeting the conditions of theorem 1.4., P being connected and $x \varepsilon X$ being arbitrary but neither a greatest nor a smallest element of X, there is a *certainty equivalent* $y \varepsilon P$ of x:

$$x \sim y.$$

Proof: Since U is continuous U(P) = u(P) is a connected set in R and hence an interval. With the proof of corollary of lemma 1.4. in mind we easily see that U(P) = U(X) holds immediately implying the assertion. ¶

The following implication of corollary of lemma 1.3. is of some interest: if the decision maker is not even informed on the distribution of the consequences, i.e. on an element of X, but on a distribution μ of X only,[1)] the utility function U is integrable with respect to the interval-σ-algebra of X:

$$\int_X U \, d\mu$$

is the expected value of the expected values of the utility u. Thus it may be spoken of a well-defined objective function in this case too, uniquely determined up to positive linear transformations as in the previous-mentioned case.

---

1) This is the problem of statistical decision theory. Additionally, the problem of decision making under uncertainty (i.e. the lack of objective probabilities) used to be attacked by some autors in that way; cf. Hart [1942, pp. 110-118].

Remark: Generally, it will be dealt either with sets of consequences having no cuts at all (discrete spaces by prop. 1.4.) - in those cases the condition (B5) is unnecessary - or with sets of consequences having cuts only (continuous sets) - in those cases the condition (B5) may be reformulated:

(B5') For all $x,y \varepsilon P$ (with $x \prec y$) and all $\alpha \varepsilon >0,1<$ there exist elements $a,b \varepsilon P$ such that

$$x \prec a \prec \alpha x + (1 - \alpha)y \prec b \prec y. -$$

Having only cuts, P is connected with respect to its interval topology; since P is connected with respect to its relative topology too (for, by (B5') P is simple in X), the proposition of the existence of a certainty equivalent holds. P being discrete the existence of certainty equivalents may not be expected in general.

1.3.1.3. *A Special Case: The Utility Function as an Algebraic Homomorphism*[1)]

The reduced set of consequences of a criterion consisting of exactly two elements, the set of probability measures of this set is provided with a natural algebraic structure enabling us to determine the utility function of the related Bernoulli-objective uniquely.

Let $P = \{x,y\}$ consist of exactly two elements where x is strictly less preferred to y. Let $M(P)$ be the set of probability measures of P. We consider x and y to be elements of $M(P)$ too. For each $\alpha \varepsilon >0,1<$ there is a binary operation on $M(P)$: $\mu \hat{\alpha} \nu = \alpha\mu + (1 - \alpha)\nu$. Applying this operation once more for $\beta \varepsilon >0,1<$ leads to $(\mu \hat{\alpha} \nu) \hat{\beta} \nu = \beta(\alpha\mu + (1 - \alpha)\nu) + (1 - \beta)\nu$ $= \mu \widehat{\alpha\beta} \nu$. Each element $\mu \varepsilon M(P)$ may be represented in the following way:

$$\mu = \alpha x + (1 - \alpha)y = x \hat{\alpha} y$$

where $\alpha = \mu(\{x\})$. Accordingly, we may define

$$\mu \dagger \nu = (x \widehat{\mu(\{x\})} y) \widehat{\nu(\{x\})} y = x \widehat{\left(\mu(\{x\}) . \nu(\{x\})\right)} y.$$

$\mu \dagger \nu$ may be interpreted to be the prospect of offering with probability $\nu(\{x\})$ x to occur with probability $\mu(\{x\})$.

---

1) For the general form of those approaches we refer to Aumann [1964, p. 219].

By † the set $M(P)$ becomes a multiplicative monoid with zero-element and unity.

Proposition 1.13. Let P consist of exactly two elements and let $M(P)$ meet the conditions of the expected-utility-theorem (theorem 1.3.). Then there is exactly one utility function u on P fulfilling the conditions of theorem 1.3. such that 1 - U is a homomorphism for the operation † and the multiplication in R. Given P as above we have $u(x)=0$ and $u(y)=1$.

The proof is obvious.

Thus, for sets of consequences consisting of two elements the objective function of a Bernoulli-objective is identified with the probability of the occurence of the greater element. As long as the elements of P are to be interpreted by *side condition h is ex post seen to be fulfilled* or by *side condition h is ex post seen not to be fulfilled* respectively, we will assume the utility function of the form given by prop. 1.13..

1.3.2. *Substantiation of the Information Requirements of the Objective-Concept*

Starting from the problem of guaranteeing liquidity in planning managerial decisions (e.g. investment and financial decisions), the following simple model may suffice to clarify the indicated constructions.

The management is to choose among n investment opportunities which may be combined and be realized at alternate levels. Denoting by x_i the level of investment opportunity i planned to be realized ($i=1,\dots,n$), there is a cash payment $a_o(x_1,\dots,x_n)$ associated with the investment vector x at the time t_o; in t_1 there is a net cash receipt denoted by $e_1(x)$. Let E be the amount of equity capital and $j=1,\dots,m$ be the opportunities of financing by debt (the associated levels planned to be realized are denoted by y_j) available in t_o. The set of vectors (x,y) will be denoted by A. A planned financing by debt y leads to a cash receipt $e_o(y)$ in t_o and to a cash payment $a_1(y)$ (redemption and interest) in t_1. No payments are assumed to occur between these dates. Then the following condition must hold:

(O) $$a_o(x) - e_o(y) \leq E.$$

Usually it is assumed that an enterprise will pay the agreed amount a_1 without any consideration of possible consequences, behaving in a way absolutely loyal to contracts; accordingly, the condition

$$(1) \qquad a_1(y) - e_1(x) \leq E - a_o(x) + e_o(y)$$

has to be fulfilled: the difference between cash payments and cash receipts must not exceed the available liquid assets. Here it is supposed that activities additional to x and y cannot be displayed.[1)]

Let t_2 be the planning horizon (where the firm is hypothetically liquidated). At t_2 cash receipts $e_2(x)$ are generated by the chosen investment opportunities and by liquidating the assets held; cash payments $a_2(y)$ are generated by repayment of debt and by payment of interest. The payment by contract must be possible in t_2 too:

$$(2) \qquad a_2(y) - e_2(x) \leq E - a_1(y) + e_1(x) - a_o(x) + e_o(y).$$

With a purely monetary objective function the problem is sufficiently described as follows:

$$\max \{Z(a_o,a_1,a_2,e_o,e_1,e_2,E) \,|\, (0) \text{ and } (1) \text{ and } (2) \text{ and } x,y \geq 0\}.$$

The functions a_i and e_i (i=0,1,2) and the parameter E may be interpreted to be points of view of an objective (under certainty). Up to now, formulating the problem has supposed complete information. Whereas there are good reasons to assume a_o,a_1,a_2 and e_o to be known in many instances, this assumption is completely unrealistic with e_1 and e_2. We therefore will assume these payments to vary stochastically: let $(\Omega,\mathcal{A},\mu)$ be the relevant probability space such that for all x the functions $e_i(x,.)$ from Ω into R are stochastic variables (i=1,2).

The side conditions (1) and (2) as well as the "max"-operator are now meaningless; it can only be stated whether or not the conditions

$$(1') \qquad (a_1(y) + a_o(x)) - (e_1(x,\omega) + e_o(y)) \leq E$$

$$(2') \qquad (a_2(y) + a_1(y) + a_o(x)) - (e_2(x,\omega) + e_1(x,\omega) + e_o(y)) \leq E$$

hold for a given $\omega\in\Omega$. Admitting actions $(x,y)\in A$ only if they meet (1') and (2') for all $\omega\in\Omega$, the problem may be given by

$$\max \{\int_\Omega Z(a_o,a_1,a_2,e_o,e_1,e_2,E)\, d\mu \,|\, (1') \text{ and } (2') \text{ for all } \omega\in\Omega\}^{2)}.$$

1) Especially neither additional financial opportunities nor a postponement may be obtained.
2) Cf. def. 1.7..

Usually this formulation is felt to be to pessimistic because it implies that for many problems there are no feasible actions at all. But admitting actions $(x,y)\varepsilon A$ partially violating condition (1') or (2') (for some $\omega\varepsilon\Omega$) makes the statement of the problem incomplete: there is no well-defined objective; for, assume condition (1') violated (e.g.) bankruptcy has taken place by assumption (because of illiquidity; by assumption there are no additional activities available in t_1), e_2 will be no longer defined, a_2 has immediately to be paid, and the objective-achievement Z cannot be determined in the indicated way; additionally the problem of how to assign objective-achievements to certain dates arises. But, if for all $\omega\varepsilon\Omega$ the objective-achievement can be determined the side condition (1') is unnecessary; then, the problem of guaranteeing the firm's liquidity has been transformed to a problem of rentability. Z becomes a more general function of x,y and ω. Similar statements hold if condition (2') is violated.

If for certain choices of (x,y,ω) the objective-achievements cannot be determined, the "danger" of such cases occurring has to be taken in and must be confronted with the "chances" of "profitable" objective-achievements.[1] Since there is no alternate information the decision maker can only construct further criteria in order to respond to this danger. In the following a natural approach is suggested:

$$\Omega^1(x,y) := \{\omega\varepsilon\Omega \mid (1') \text{ does not hold and Z cannot be determined}\}$$

$$\Omega^2(x,y) := \{\omega\varepsilon\Omega \mid (2') \text{ does not hold and Z cannot be determined}\}$$

We are now prepared to define three criteria:

a) The set of points of view for the first criterion is given by

$$H_1 := \{E\ ;\ a_i,e_i : \bigcup_{(x,y)\varepsilon A} \{(x,y)\}\times\Omega(x,y) \rightarrow R \mid i=0,1,2\}$$

where $\Omega(x,y) := \Omega\setminus(\Omega^1(x,y)\cup\Omega^2(x,y))$. The criterion function is given by

$$Z(a_o,a_1,a_2,e_o,e_1,e_2,E).$$

b) The criterion functions of the remaining two criteria may directly be written as

$$Z_i = 0_i \text{ for } \omega\varepsilon\Omega^i(x,y) \text{ and } Z_i = 1 \text{ elsewhere} \qquad (i=1,2).$$

1) The previous-mentioned method of excluding actions from consideration which generate such a danger at all turns out to be a special case of the approach here indicated.

The chosen formulation of the problem which originally had one objective under certainty only has , under uncertainty, led to a problem with three *partial* objectives (criteria).

Of course it is unrealistic to assume that no further activities are available after t_o, since in most cases there are facilities for obtaining a moratorium or for opening up additional financial sources; these facilities might have not been considered before because of a high associated level of costs or because of additional non-monetary disadvantages: e.g. the period of payment of credits allowed by suppliers may often be overstepped without involving bankruptcy; it may seem more profitable to creditors to agree upon a moratorium rather than to participate in the bankruptcy's estate because of an evaluation of the long -term evolution of the firm's capital assets; the firm may sell assets (e.g. receivables) prior to their maturing, i.e. before the assets in question will normally become liquid (factoring). The first two operations being to shift the *dates of cash payments* to future periods, the latter is to remove *dates of cash receipts* from future to present time. Such removements in general involve disadvantages of a monetary or non-monetary kind: higher interest rates, commissions, discounts; a reducing of the debt-financing opportunities in the long run, a damage of the firm's goodwill etc.. Many of these consequences cannot immediately be transformed to monetary objective-achievements; in appropriate instances they have to be subjected to separate partial objectives.

The indicated *balancing activities*[1)] can be taken into account within the framework of flexible planning models only. Let Ω_1 be the set of environmental responses in t_1 influencing the value of e_1: then e_1 is defined for all choices of (x,ω) where $\omega\varepsilon\Omega_1$. Let X be the set of combinations (x,y,ω) fulfilling the inequality (1'); let $\bar{X}$ be the set of combinations which do not meet (1'). We will assume that there are no additional activities available after $v\varepsilon X$ having occured; the final states in t_2 will be assumed to be given by $X\times\Omega_2$ where Ω_2 describes the set of environmental responses to combinations of X. For each action (x,y) the set $\{(\omega_1,\omega_2)\varepsilon\Omega_1\times\Omega_2|(x,y,\omega_1)\varepsilon X\}$ is supposed to be a sample space.

For each $v\varepsilon\bar{X}$ there is a set Z_v of possible (considered) activities compensating the bottleneck of liquidity and revising the original action. Naturally $Z_v = \emptyset$ may occur: there are cases no balancing activities

1) *Ausgleichsaktivitäten* in the sense of Riemenschnitter [1972, pp. 65- - 66].

exist in.[1] Furthermore, to each $z \varepsilon Z_v$ there is a set $\Omega(z)$ of possible environmental responses such that the final states may be described by (v,z,ω) with $z \varepsilon Z_v$ and $\omega \varepsilon \Omega(z)$. Every map $s : Y \rightarrow \bigcup_{v \varepsilon \bar{X}} Z_v$ generates a strategy[2] where $Y := \{v \varepsilon \bar{X} | Z_v \neq \emptyset\}$ and $s(v) \varepsilon Z_v$ for all $v \varepsilon Y$: to each (x,y) the map s assigns the associated activity to be chosen when ω occurs. For each activity (x,y) and each mapping s of the mentioned kind the set $\{(\omega_1,\omega_2) | (x,y,\omega_1) \varepsilon \bar{X}$ and $\omega_2 \varepsilon \Omega(s(x,y,\omega_1))\}$ is supposed to be a sample space. Formally defining $Z_v := \{(x,y) | v=(x,y,\omega)\}$ for each $v \varepsilon X$, the maps s may be extended to $X \cup Y$. Now, the actions of the model are given by pairs $((x,y),s)$ where (x,y) symbolizes the initial activity and s the balancing and revising strategy. Thus the flexible planning model is reduced to the structure of a decision situation under uncertainty disregarding the description of the final states by aspects - naturally the sample spaces have still to be provided with probability measures.

Two facts show that nothing fundamental has been changed compared with the initial statement of the problem: on the one hand, the same problems arise when $Z_v = \emptyset$ holds, problems of evaluating the case of illiquidity; on the other hand, this evaluation must probably be carried out in t_2 too: it may be thought the decision maker is not sufficiently informed of final states in t_2 to determine the associated objective-achievement; here also bottlenecks of liquidity may occur, the decision maker cannot say a priori whether or not they can be compensated.

The depicted problems justify the information assumptions of our definition of partial objectives. At the same time it becomes clear that decisions under uncertainty in general are decisions with multiple criteria, for, at least the dual criterion or a decomposition of the dual criterion in several (0,1) criteria has to be considered as soon as final states may occur which cannot be evaluated. At last it is seen that only actions may be excluded which cannot be carried out with probability 1 or which lead to unfeasible final states with probability 1.

---

1) These cases are called *dead knots (tote Knoten)* by Riemenschnitter [1972, p. 72].

2) In flexible planning models the *actions* in the sense of stiff planning models are replaced by *strategies*. A strategy is a sequence of actions in the sense of stiff planning, each of them related to certain events (Riemenschnitter [1972, p. 47]).

1.3.3. An Example

A further illustration of the introduced concepts will be carried out by the mentioned example of a flexible planning model of a decision on financial and investment opportunities. Let x symbolize an investment decision leading in t_o to a cash payment $a_o(x)$, y a financial decision leading in t_o to a cash receipt $e_o(y)$. If necessary both decisions will be revised if they turn out to be revisable. Let A_o be the set of considered pairs (x,y). Ω_1 is the set of environmental responses to decisions of A_o. $A_o \times \Omega_1$ is the set of realizable states in t_1. The states (x,y,ω) lead to a cash receipt $e_1(x,y,\omega)$ if no revision will be carried out, and to a cash payment $a_1(x,y,\omega)$ if no revision will be carried out and if there will be a sufficiently large amount of liquid assets in order to pay $a_1(x,y,\omega)$: this is the case $(x,y,\omega)\varepsilon X$.

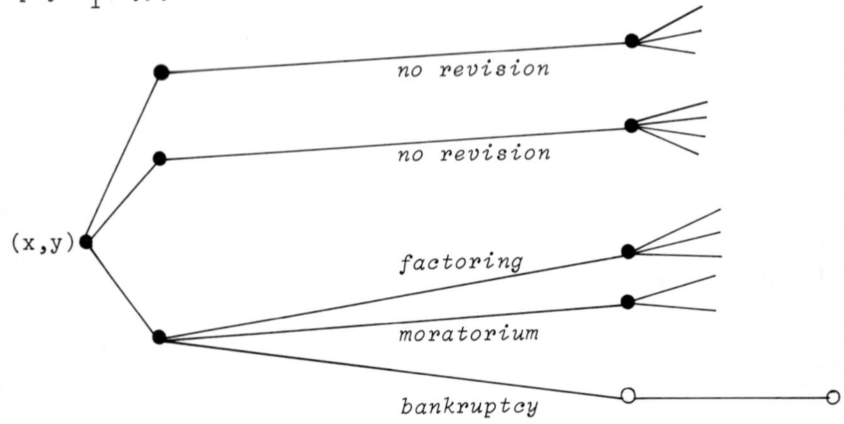

If (x,y,ω) belongs to $\bar{X}$, i.e. a revision becomes necessary, the cash payments and the cash receipts in t_1 depend on the revision carried out in t_1: if $z \varepsilon Z_{(x,y,\omega)}$ stands for, say, factoring an additional cash receipt will be generated; if z stands for obtaining a moratorium from the creditors a part of the planned payments will be left undone. If there is no balancing activity at all, probably no statements can be made on the true received net cash receipts. Thus for all (x,y,ω) and all $z\varepsilon$ $\varepsilon Z_{(x,y,\omega)}$ cash receipts $e_1(x,y,\omega,z)$ and cash payments $a_1(x,y,\omega,z)$ in t_1 are determined; with the above-mentioned formal assumption in mind we may define functions e_1 and a_1 on the set of states (x,y,ω,z).

By the environmental response ω' the final state (x,y,ω,z,ω') is generated. This final state involves a cash receipt $e_2(x,y,\omega,z,\omega')$ in t_2 and the obligation to pay $a_2(x,y,\omega,z,\omega')$. If this amount can be paid, no problem arises; if it cannot be paid, the planning depends on whether the decision maker is informed which the opportunities are to compensate the financial gap. Under appropriate instances this state must be treated as a dead knot.[1] To facilitate the discussion this will be assumed in the following.

For an application of the concept of a criterion to the chosen example we confine ourselves to the mentioned decision situation in order to avoid complications; thus we omit to investigate a whole class of situations; such an investigation was carried out by an analysis of situations under certainty.

We may suppress a description of aspects because all information has been already given in the form of points of view. If the cash payments really paid in the situation in question are denoted by a_2' we may use three functions to characterize points of view: e_2, a_2, a_2'.[2] For a monetary criterion the points of view e_2 and a_2' are sufficient;[3] the criterion function may be given by

$$Z_o := e_2(x,y,.,z,.) - a_2'(x,y,.,z,.).$$

If a bottleneck of liquidity occurs a_2' is unknown, the function Z_o is defined only on the subspace the function $a_2'(x,y,.,z,.)$ is defined on. The point of view a_2 may be used to define the dual criterion:

$$Z_1(x,y,\omega_1,z,\omega_2) := \begin{cases} 1 & \text{for } e_2 \geq a_2 \\ 0 & \text{elsewhere.} \end{cases}$$

1.4. *Partial Objectives and Managerial Decisions*

It is an unchallenged fact that managerial decisions in reality are made

1) This makes sense if it is assumed that the planning horizon is fixed at that point, and no additional information is available after wards. For this assumption we refer to Krümmel [1964, p. 194].
2) a_2' contrasts with cash payments which were to be paid by the original contract.
3) Here it is supposed that there is a decision variable x_1 associated with the holding of liquid assets.

under consideration of several criteria.[1] On the one hand, the existence of multiple criteria is based on the fact that in reality it cannot be spoken of as decision *unity*: decisions within organizations are in fact results of balancing the individual efforts; the firm's decisions are subjected to several objectives because several persons with differing individual objectives have a (direct or indirect) influence on the decisions. This mainly results in *interpersonal* objective-conflicts. An important example is the conflict among the objectives of the shareholders and of the management of a company.[2]

On the other hand, however, the fictitious all-deciding manager who owns the firm by himself - a fiction the science of industrial management is widely based on - does not even guarantee that only one objective is relevant for all decisions. This is a consequence of at least the following two facts:[3]

1. Real managerial decisions are characterized by incomplete information on the parameters of the decision situation.
2. The personality structure of real individua is determined by various different motivations.

This results in *intrapersonal* objective-conflicts.

Ad 1. By incomplete information objective-conflicts may be generated in two ways:

a) Defining and discussing our concept of partial objectives we have remarked that final states may occur, such that, chosen a partial objective the given information is not sufficient to determine the objective-achievements. Accordingly, the dual criterion has to be considered additionally, unless these final states are treated by any other separate partial objective. Choosing an arbitrary partial objective, one necessarily has to take in a further criterion because of the depicted lack of information.

---

1) We give the following references: Heinen [1966], Schmidt-Sudhoff [1967], Bidlingmayer [1968]. Based on empirical investigations Kaplan, Dirlam and Lanzilotti [1958] and Raia [1965] have catalogued managerial objectives while at the same time pursuing real decisions.
2) Another important example consists of the conflict among the objectives of the heads of different departments. Planning the product range the production manager oftenly intends to reduce the number of different products, whereas the head of the marketing department looks for chances to sell a greater amount of products because of a manifold assortment of products he can supply, cf. Gutenberg [1968, Vol.I, pp. 153-154].
3) Dealing with side conditions of type B we have seen another reason.

b) The lack of information has a temporary dimension too: economical activities can be planned for a limited period of time only. The limit of time in question is determined by the temporary threshold of information: for dates after this point there is no sufficient relevant information at all. If all information at hand is to be used the planning horizon has to be fixed at this point.[1] Partial objectives implicitly depend on the planning horizons of the decision situations concerned, as the sets of considered actions and the sets of environmental responses are related to the planning horizon.

This temporary dimension of criteria may lead to intrapersonal conflicts along the following line: planning with a planning horizon of ten years by means of profit-maximization (however defined) the manager might fear the pursuing of this objective alone could reduce the long-term profit potential, for it leads to strategies which settle the firm's structure in a very one-sided way. This fear may be ruled out by treating diversification (however defined) as an additional criterion. In so doing the manager might think performances with respect to this objective to have a positive influence on the long-term profit potential.[2]

Arguing this way indicates an approach to solve such objective-conflicts: to assume a superordinate objective the decision maker cannot directly formulate in an operational form. Then the problem consists of the search for an operational representation of the superordinate objective, at least as far as is necessary to find an "optimal" decision.

Ad 2. By ensured findings of psychology the actions of real persons are determined by a complex structure of motivations.[3] This should be accounted for when constructing a normative theory if this needs to be consulted in order to improve real decisions. But in such a theory there is no place for the concept of a decision maker, who surrenders himself unreflecting to his incited motivations, i.e. such a theory must require the decision maker to transform his motivations to methods of evaluation generating consistent orderings, e.g. by criteria in the sense of our definition; this will be assumed later on.

The following deals with the question of how the decision maker solves (is to solve if we are more precise) the generated intrapersonal objective-conflicts, i.e. how he chooses a certain action faced with his

1) Cf. Krümmel [1964, p. 194].
2) "... it is possible to measure certain present characteristics of firms which are likely to influence their profit potential", White [1960, p. 186].
3) As an example from an extensive literature we refer to Wundt [1965].

conflicting motivations.

In these notes we will treat intrapersonal conflicts only; certainly, interpersonal conflicts require different methods from those used to solve intrapersonal conflicts, for various types of these conflicts may occur differing by the organizational structure within which they take place, and differing by the capability of the concerned persons to prevail.

At first we still have to declare the way we will deal with the intrapersonal objective-conflicts - in the following simply referred to as 'objective-conflicts'. - Basically there are two different approaches with different intentions: a theory of economic decisions may try to develop instruments to forecast the behavior of economic actors. Forecasting is usually based on a description of the real behavior rather than on a construction of a logically consistent behavior. Models describing the real behavior[1] are of some importance to forecast the consequences of actions the decision maker may choose. But assuming a given information level we are no longer interested in such models; we are interested in the second approach to the problem of objective-conflicts: we want to aid the decision maker to improve his decisions, we want to develop methods to make decisions *rational*.

2. *Formal Statement of the Problem*

2.1. *Complete Systems of Objectives*

We start with a class $\mathfrak{A}$ of decision situations under uncertainty. Let (H,Q,g) be the representation of a criterion for $\mathfrak{A}$. We consider a decision situation $\Gamma=(A,\langle\Omega,\mathcal{A},\mu\rangle_A)\varepsilon\,\mathfrak{A}$. For every $a\varepsilon A$ the set $X_{\Gamma,a}=\{\omega\varepsilon\Omega_a|(\Gamma,a,\omega)\varepsilon\bigcap_{(T,\gamma)\varepsilon H}E_{T(\Gamma)}\}$ is defined by (v) of def. 1.6.; this set is an element of $\mathcal{A}_a$, it describes the set of environmental responses to a, whose associated final states can be evaluated by the mentioned criterion.

Definition 2.1. A *system of objectives for* $\mathfrak{A}$ is a finite set of partial objectives for $\mathfrak{A}$. A system of objectives is called complete if the following conditions hold:

(i) If by $Y_{\Gamma,a}$ is denoted the intersection of the sets

---

1) For such models we refer to Heinen [1966], Schmidt-Sudhoff [1967] and Bidlingmayer [1968].

$X_{\Gamma,a}$ related to the objectives concerned we require for all $a\varepsilon A$: $\mu_a(Y_{\Gamma,a}) \neq 0$ for all situations $\Gamma\varepsilon\mathfrak{A}$.

(ii) If by $\bar{Y}_{\Gamma,a}$ is denoted the union of the sets $Y_{\Gamma,a}$ we we require $\mu_a(\bar{Y}_{\Gamma,a}) = 1$ for all $a\varepsilon A$ and $\Gamma\varepsilon\mathfrak{A}$.

Condition (i) states that almost always there is no action generating final states which cannot be evaluated by at least one criterion of the system concerned. Condition (ii) requires that each final state in $\mathfrak{A}$ can almost certainly be evaluated by at least one criterion of the system. By (ii) is ensured that almost certainly all occuring final states have some influence on the decision.

By condition (v) of def. 1.6. the definition of completeness is independent of the particular representation of the criteria; completeness is therefore well-defined.

Condition (i) may be enforced by ruling out actions which do not meet it; this seems reasonable as long as all partial objectives of the system are relevant as far as the decision maker is concerned. Condition (ii) seems reasonable too, for the probability of making a misjudgment by ruling out final states from consideration should equal 0. This condition may be met too if the related dual criterion is added to the system.

Proposition 2.1. Let (H,Q,g) be a V-criterion for $\mathfrak{A}$. For each decision situation $\Gamma=(A,<\Omega,\mathcal{A},\mu>_A)$ in $\mathfrak{A}$ we may construct a function (the criterion function)

$$Z^\Gamma : \Phi(\Gamma) \rightarrow R\cup\{0_\Gamma\}$$

such that the pair $((Z^\Gamma)^{-1}(R),Z^\Gamma)$ is (slightly transformed) an object of the category V, and this object is uniquely determined up to isomorphisms of V by the chosen objective. The same notation is used as in 1.3..

Proof: For all points of view $(T,\gamma)\varepsilon H$ and all decision situations Γ the following diagram is commutative:

$$\begin{array}{ccccccc}
\bigcap\limits_{(T,\gamma)\varepsilon H} \mathfrak{A}(T) & \xrightarrow{inj_\Gamma} & E_{T(\Gamma)} & \xrightarrow{s_{T(\Gamma)}} & \prod\limits_{n\varepsilon T(\Gamma)} E(n)/G(n) & \xrightarrow{i} & \bar{\mathfrak{A}}_T \\
\downarrow{\scriptstyle inj_\Gamma} & & \downarrow{\scriptstyle i} & & \downarrow{\scriptstyle i} & & \downarrow{\scriptstyle p_\gamma} \\
\mathfrak{A}(T) & \xrightarrow{id} & \mathfrak{A}(T) & \xrightarrow{\Delta_T} & \bar{\mathfrak{A}}_T & \xrightarrow{p_\gamma} & \mathfrak{A}_{T_\gamma}
\end{array}$$

p_γ denotes the canonical projection and i the respective inclusion maps. There is one and only one mapping ψ_H making the following diagram commutative for all $(T,\gamma)\varepsilon H$:

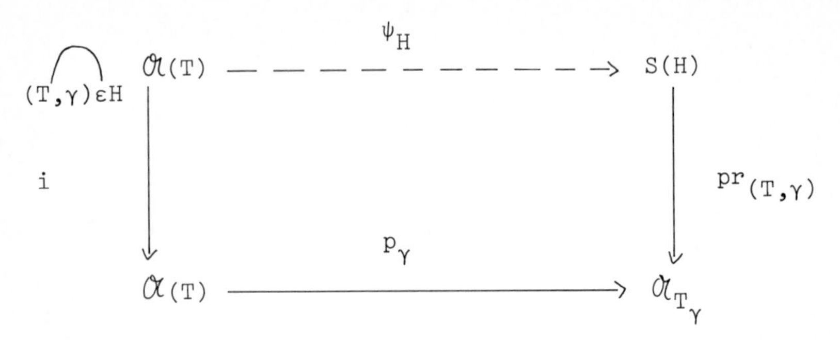

ψ_H is the result function of the criterion (cf. def. 1.6.). We define

$$z^\Gamma(a,\omega) = \begin{cases} g\cdot\pi\cdot\psi_H\cdot inj_\Gamma(a,\omega) & \text{if } inj_\Gamma(a,\omega)\varepsilon \bigcap_{(T,\gamma)\varepsilon H} \mathfrak{A}(T) \\ 0_\Gamma & \text{elsewhere.} \end{cases}$$

The assertion concerning Z^Γ is an immediate consequence of condition (vi) and condition (vii) of def. 1.6.. ¶

Corollary If (H,Q,g) is a Bernoulli-objective (with transformation group D) and if $\mu_a(X_{\Gamma,a}(H,Q,g)) \neq 0$ holds for all $a\varepsilon A$ given a $\Gamma\varepsilon\mathfrak{A}$, then there is a mapping

$$z^\Gamma : A \rightarrow R$$

uniquely determined up to transformations belonging to D.

Proof: Z^Γ given by prop. 2.1. by definition 1.7. the expression

$$z^\Gamma(a) = \int_{X_{\Gamma,a}(H,Q,g)} Z^\Gamma(a,.)\ d(\mu_a|X_{\Gamma,a}(H,Q,g))$$

can be formed. The uniqueness is an immediate consequence of the linearity of the integral operator and of proposition 2.1.. ¶

2.2. *Criteria Vectors*

Given a complete system of objectives $\{(H_i,Q_i,g_i)|i=1,\dots,n\}$ for $\mathfrak{A}$ proposition 2.1. generates a mapping

$$Z^{\Gamma} : \Phi(\Gamma) \rightarrow \prod_{i=1}^{n}(R \cup O_{\Gamma}^{i})$$

for each $\Gamma=(A,<\Omega,\mathcal{A},\mu>_A)$. For this mapping uniqueness statements hold which depend on the related categories V_i. For each $a\varepsilon A$ there is naturally defined a probability space on Im $Z^{\Gamma}(a,.)$. Z^{Γ} is a real vector function on the subset

$$\bigcup_{a\varepsilon A} \{a\}\times Y_{\Gamma,a}$$

of $\Phi(\Gamma)$. Throughout the following we will assume that for partial objectives of a complete system of objectives condition (iii) of def. 1.7. holds, i.e. that the criterion functions (vitually) are measureable.

The problem of objective-conflicts under uncertainty may now be stated as follows:

> *Choose an action* $a\varepsilon A$ *such that the resulting distribution* $Z^{\Gamma}(a,.)$ *of the criterion values becomes 'best'!*

It remains to make clear how the 'best' distribution of criterion values is defined. It has been suggested by Dinkelbach and Isermann [1973] that the approaches of the decision theory under uncertainty may be applied to the criteria Z_i seperately; no matter which solution approach is applied (min-max-principle, Wald-criterion, Bernoulli-principle etc.), in all cases the distribution $Z^{\Gamma}(a,.)$ is replaced by a value $z^{\Gamma}(a)$ (in some sense a 'certainty equivalent') and the problem is transformed to the vector maximum form:

$$z^{\Gamma} : A \rightarrow R^n$$

By constructing isolated certainty equivalents a probably decisive truncation of information may happen: information, e.g. provided by the covariances of the criteria, is ruled out; for n=2 actions a_1 and a_2 may have the same combination of isolated certainty equivalents, whereas with a_1 a correlation of 1 and at the same time with a_2 a correlation of -1 between the criteria may be associated.[1)] According to the decision maker's attitudes a_1 is prefered to a_2 (if the decision maker is in some sense a risk-lover) or vice versa (if the decision maker is in some sense a risk-averter), though both actions are characterized by the same vector $z^{\Gamma}(a_1)=z^{\Gamma}(a_2)$.

---

1) With respect to competing partial objectives prevailingly negative correlations are expected to occur.

As well as the interdependent structure of risks between the components of a portfolio and the risk-destroying property of composing a portfolio is explained even by a simple consideration of the expected value and of the variance of the portfolio,[1] an insight into the specific problems of multiple criteria decision making under uncertainty may be obtained by a similar approach.[2] In order to reflect the levels of expected objective-achievements the function $z^\Gamma : A \to R^n$ consisting of the expected values of the random variables $Z_i^\Gamma(a,.)$ may be considered (supposed to exist). We already remarked that this function neglects the risks or chances respectively of two partial objectives to reach high or low levels at the same time or to contrarotate, i.e. it neglects the stochastic interdependence of the criteria.

For a tentative and in general an incomplete comprehension of these phenomena the structure of covariances of the system may be used:

$$R_{ij}^\Gamma(a) := \operatorname{cov}\left(Z_i^\Gamma(a,.), Z_j^\Gamma(a,.)\right)$$

$R_{ij}^\Gamma(a)$ may be interpreted as the *partial risk between the criteria i and j with respect to the action* a.[3] Approximately, the problem may be characterized by a mapping $p^\Gamma : A \to R^n \times R^{n^2}$ where $p_i^\Gamma(a) = z_i^\Gamma(a)$ for $i=1,\ldots,n$ and $p_{n.i+j}^\Gamma(a) = R_{ij}^\Gamma(a)$ for $i,j=1,\ldots,n$.[4] Formally, this is a problem under certainty. We do not go into further details on this approach now; the specific problems of multiple criteria under uncertainty may be made clearer by this suggestion, but to reach a more general and theoretically satisfying solution more information is needed.

A decision behaviour which follows the vector of probability distributions of the partial objectives only will be called *indifference against the objective-risk*; if the matrix $R^\Gamma(a)$ of partial risks is not diagonal for at least one action a we will speak of a decision situation with *objective-risk*.

If all V_i are D-categories the matrices $K^\Gamma(a)$ of intercorrelations ($K_{ij}^\Gamma(a)$

---

1) Cf. Markowitz [1959].
2) Discussing a quadratic model will make clear the affinity of both methods (section 3.2.).
3) This notion must be handled with some care, for it is heavily based on the chosen representation of the objective-system by the corresponding criteria functions; but these functions are unique up to certain transformations in R only.
4) In order to have real criteria functions we must apply utility functions of the form $u_{ij}=id$ (or $u_{ij}=-id$) asserting an increasing R_{ij}^Γ to be preferable (or not preferable).

$= R^{\Gamma}_{ij}(a)(var(Z^{\Gamma}_i(a,.).var(Z^{\Gamma}_j(a,.))^{-1/2})$ may be subjected to a factor analysis leading to a system of "independent criteria";[1] this would meet the decision maker's need for a rational clearing up of the system of objectives and could help the decision maker to construct his objectives and to make them operational.

2.3. *The Treatment of the Problem on Principle*

Given a complete system of objectives $\bar{Z} := \{(H_i,Q_i,g_i) | i=1,\dots,n\}$ for the class $\mathfrak{A}$ with the criterion functions Z_i we proceed in the same way as solving the problem of a single objective under uncertainty: the vector function $Z = (Z_1,\dots,Z_n)$ induces an equivalence relation γ on $\mathfrak{A}^*$[2] which is independent of the chosen representation. This relation γ together with the relevant aspects given by the system of objectives concerned may be understood as a point of view under uncertainty. The equivalence classes modulo γ may be identified with a subset of $R(\mathfrak{A}) := \prod_i (R \cup \bigcup_{\Gamma \in \mathfrak{A}} \{0^i_\Gamma\})$ by virtue of the function Z; this identification naturally is with respect to the special representation of the system of objectives and thus unique only up to certain transformations in R^n determined by the underlying categories V_i $(i=1,\dots,n)$. By a *superobjective* for the pair $(\mathfrak{A},\bar{Z})$ (in the following this pair is referred to as a problem of *multi-objective decision making under uncertainty*) is meant a Bernoulli-objective (H,Q,g) for the class $\mathfrak{A}$ with $H = \{\gamma\}$. The set of consequences may be identified with $\operatorname{Im} Z \subset R(\mathfrak{A})$, Q may be assumed to be given on Im Z (in this form it depends on the chosen representation) as well as $g : \operatorname{Im} Z \to R$. We require the following conditions to hold for a superobjective:

(i) Q is weakly monotone[3] with respect to the preordering $\leq$ induced on $R(\mathfrak{A})$ by the natural preordering of R^n in the following way: $x,y \in R(\mathfrak{A})$

---

1) In this case the matrices $K^{\Gamma}_{ij}(a)$ are uniquely determined by the objective-system concerned; for the method of factor analysis we refer to Überla [1968] and the references given there.

2) In the way indicated in footnote of p. 8.

3) Let M be a set provided with a preordering $\leq$ (i.e. a reflexive and transitive binary relation on M); a binary relation Q on M is called *weakly monotone* with respect to $\leq$ if and only if Q $\leq$. By the *natural preordering* of R^n we mean the following relation: $x \leq y \Leftrightarrow x_i \leq y_i$ for all $i=1,\dots,n$. Frequently we will use the following notation for $x,y \in R^n$: $x<y \Leftrightarrow x \leq y$ and $x \neq y$; $x \ll y \Leftrightarrow x_i < y_i$ for all $i=1,\dots,n$.

$x \leq y \Longleftrightarrow$ (1) and (2): (1) $x_i = 0^i_\Gamma$ iff $y_i = 0^i_\Gamma$

(2) $x_j \leq y_j$ for all j with $x_j \in R$

(ii) The trace of Q on R^n is continuous with respect to the relative topology of Im $Z \cap R^n$.

Conditions (i) and (ii) imply that g is continuous and monotone increasing on Im $Z \cap R^n$. In most cases we replace Im Z by $B(Z) = \text{Im } Z \cup \overline{cv(\text{Im } Z \cap R^n)}$ and we will assume that Q and g are the traces of corresponding objects $\bar{Q}$ and $\bar{g}$ defined on B(Z). Replacing Im Z by B(Z) corresponds to an addition of the expected values of all probability distributions on Im $Z \cap R^n$ (in a theoretic game context: turning from pure to mixed strategies).

Along this line the problem has been formally solved; in each situation $\Gamma \varepsilon \mathfrak{A}$ has to be in accordance with the superobjective function

(†) $\max\{\int g \cdot Z^\Gamma(a,.)\, d\mu_a | a \varepsilon A\}$.

But here we meet with the real problem: if the function g were at hand, a problem of multiple objectives would not really exist. The problem consists in the missing knowledge of g; consequently, we have to develop methods to discover the function g. This will be dealt with in the following. Special difficulties with respect to the treatment of the function g occur at points of B(Z) where $x_j = 0^i_\Gamma$ holds for some i and Γ. The easiest way of treating such points is to exclude them from further consideration by using the conditional expected value in (†) with respect to the subspace $Y_{\Gamma,a}$ (by assumption of definition 2.1. we have $\mu_a(Y_{\Gamma,a}) \neq 0$); this implies that final states leading to points of the mentioned kind are on principle treated on average as "good" as the remaining states generated by the same action. For one thing this approach has practical advantages, for another it is not unreasonable as long as all partial objectives are relevant and all final states the decision maker is not completely informed on may be treated with the same reasons as favorable and as unfavorable; but this argument neglects the fact that different final states of this kind might be differently looked upon, according to which levels the remaining partial objectives are achieved with regard to these final states. However, we will proceed as indicated; in some cases where preference relation Q meets the condition (INC) (see p. 72), the evaluation of the elements 0^i_Γ can easily be integrated into the procedure concerned.

At first the following remarks on our general solution approach seem necessary: in accordance with the solution for the problem of a single criterion[1] and according to our program to prescribe standards concerning a "rational" decision behavior in view of several criteria[2] we require the decision maker to behave "rationally" in the sense of the Bernoulli-principle.[3] This seems a reasonable working hypothesis all the more since this principle is generally accepted as a basis for decision models under uncertainty,[4] even if it has frequently to be subjected to rigorous simplifications,[5] for in general the special pattern of the utility function is hardly available.[6] Additionally, the chosen approach allows treating the previously indicated objective-risk problems in full generality. On the other hand we will see that just this property involves unfavourable consequences for other components of the problem of multi-criteria decision making. In special cases where the operationality is not exposed to danger, we will revise this general solution approach.

Stating the problem in the form $Z^{\Gamma} : \bigcup_{a \varepsilon A} \{a\} \times Y_{\Gamma,a} \rightarrow R^n$ is formally very similar to the vector maximum problem; thus it may be suggested to apply methods for solving those problems: in place of the set of actions (under certainty) here will be dealt with the set of consequences of a stochastic decision situation. This suggestion proves right only in partial; in a short excursus we will set forth the vector maximum problem (interpreted as the problem of multi-criteria decision making under certainty) and will discuss the applicability of the methods provided by the literture with respect to our questioning. It will turn out that only the recently developed approaches for solving the vector maximum problem are on principle unsuitable to a general treatment of the problems under uncertainty - as long as in accordance with the Bernoulli-principle, which literature forms the basis for dealing with uncertainty - because they do not start from an appropriate superobjective function. But as far as we can see, there are up to now no satisfying approaches to determine superobjective functions.

---

1) Cf. definition 1.7. of these notes.
2) Cf. p. 50
3) The equivalence of the assumptions of the Bernoulli-principle with our objective-concept has been shown in an excursus.
4) Krümmel [1969, p. 73, footnote 9].
5) See, e.g. the application in the form of the (μ,σ)-principle within the portfolio selection theory, Markowitz [1957].
6) Mosteller and Nogee [1951] have done considerable work on measuring these functions.

2.4. *Excursus: The Vector Maximum Problem*

2.4.1. *The Concept of a Solution. Some Results*

The problem of multi-objective decision making under certainty may be formally described by the vector maximum problem.[1] Given a system $Z := \{(H_i, Q_i, g_i) \mid i=1,\dots,n\}$ of objectives under certainty for the class $\mathfrak{A}$, there is for each $i=1,\dots,n$ and each set of actions $A \varepsilon \mathfrak{A}$ the objective function $Z_i^A : A \rightarrow R$ determined by virtue of[2] $Z_i^A = g_i \cdot \pi_i \cdot \psi_{H_i}(A,.)$; according to the category V_i associated with the objective i this objective function is unique only up to certain transformations in R. We are led to the vector maximum problem by composing $Z^A := (Z_1^A, \dots, Z_i^A, \dots, Z_n^A)$. If T_i is the set of transformations associated with objective i the function Z^A is uniquely determined up to elements of $T = \prod_i T_i$. We are now ready to state what is meant by a solution of the vector maximum problem.

Definition 2.2. By a vector maximum problem of dimension n we mean a pair $(\mathfrak{A}, Z)$ consisting of a class $\mathfrak{A}$ of decision situations under certainty and a system Z of n objectives for $\mathfrak{A}$. By a *solution* of the vector maximum problem $(\mathfrak{A}, Z)$ we understand a mapping

$$S_Z : \mathfrak{A} \rightarrow \mathfrak{P}\left(\bigcup_{A \varepsilon \mathfrak{A}} A\right)$$

with $S_Z(A) \subset A$ for all $A \varepsilon \mathfrak{A}$. The set $S_Z(A)$ will be referred to as the *set of solutions of the problem* (A,Z).

In the following we will introduce some postulates restricting the general solution concept to that of a "rational" solution.

Efficiency Postulate[3]

(EP) For all $A \varepsilon \mathfrak{A}$, all $a \varepsilon A$ and all $b \varepsilon S_Z(A)$ we have

$$Z^A(a) \geq Z^A(b) \Rightarrow Z^A(a) = Z^A(b).$$

(EP) is easily seen to be independent of the special objective functions chosen above. (EP) requires an action whose vector of objective-achievements is dominated by the corresponding vector of another action not to belong to the set of solutions; this is obviously a reasonable require-

1) The vector maximum problem was first mentioned by Kuhn and Tucker [1951].
2) Cf. definition 1.4. of these notes.
3) This condition is referred to as "functional efficiency" by Charnes and Cooper [1967, Vol. I, p. 321].

ment.

The following requirement is always implicitly postulated in treating the vector maximum problem:

Postulate of the Exclusive Relevance of the Objectives

(PERO) For all $A \varepsilon \mathcal{A}$ and all $a, b \varepsilon A$ we have

$$a \varepsilon S_Z(A) \text{ and } Z^A(a) = Z^A(b) \Rightarrow b \varepsilon S_Z(A).$$

(PERO) requires that, as far as the search for "best" actions is concerned, actions differ only if they are represented by different vectors of objective-achievements.

Definition 2.3. Let S_Z and S_Z' be solutions for the vector maximum problem $(\mathcal{A}, Z)$. S_Z will be called *finer* than S_Z' (or S_Z' will be called *coarser* than S_Z respectively) if for all $A \varepsilon \mathcal{A}$ $S_Z(A) \subset S_Z'(A)$ holds.

Definition 2.4. By the *complete solution*[1] we mean the solution S_Z^c defined by

$$S_Z^c(A) := \{a \varepsilon A \mid \neg \bigvee_{b \varepsilon A} Z^A(b) > Z^A(a)\}$$

for all $A \varepsilon \mathcal{A}$.

Proposition 2.2. The complete solution of a vector maximum problem is the coarsest solution fulfilling (EP) and (PERO).

The proof is obvious.

We add a further requirement which seems natural too, but it has interesting structural consequences as we will see.

Postulate of the Independence of Irrelevant Alternatives[2]

(PIIA) Suppose for all $A, A' \varepsilon \mathcal{A}$ with $A' \subset A$ and for all $a \varepsilon A'$ $Z^A(a) = Z^{A'}(a)$ then we have

$$S_Z(A) \cap A' = \begin{cases} \emptyset \\ S_Z(A') \end{cases}$$

(PIIA) requires that "best" actions for A remain "best" with respect to

1) The concept of the complete solution is due to Dinkelbach [1971, p. 2].
2) See the related "condition 3" for a social welfare function given by Arrow [1951, p. 27] or the "axiom 4" given by Harsanyi [1959, p. 330] in order to treat axiomatically Nash's solution for a cooperative n-person game.

a reduced set A' of actions, unless they are ruled out by reducing, and conversely that *all* "best" actions for A' are "best" for A, if there are any "best" actions of A' being "best" actions of A at the same time.

Postulate (PIIA) implicitly gives an interpretation to the abstract solution concept: all elements of the set of solutions are treated as of equal desirability. By means of this interpretation (PIIA) is in some sense a consistency assumption: defining for each $A\varepsilon\mathfrak{A}$

$$a \sim_A b \quad \Longleftrightarrow \quad a\varepsilon S_Z(A) \text{ and } b\varepsilon S_Z(A)$$

and

$$a \prec_A b \qquad a\notin S_Z(A) \text{ and } b\varepsilon S_Z(A),$$

the postulate (PIIA) may be reformulated:

$$a,b\varepsilon A': \quad a \prec_A b \Rightarrow a \prec_{A'} b$$

$$a \sim_A b \Rightarrow a \sim_{A'} b.$$

A remarkable fact is that the complete solution does not meet this requirement; this is seen to be reasonable as the complete solution does not bring the objective-conflict to a head, but only allows it to become evident; not all efficient vectors of objective-achievements are of the same desirability, but a priori they all have the same chance to be chosen, unless there is something known about the decision maker's specific preferences.

As an interesting structural consequence of postulate (PIIA) we prove the following

Theorem 2.1. Suppose for all $A\varepsilon\mathfrak{A}$ and all $a,b,c\varepsilon A$ we have $\{a,b,c\}\varepsilon\mathfrak{A}$ and $Z^A(x) = Z^{\{a,b,c\}}(x)$ for all $x=a,b,c$. Let S_Z be a solution fulfilling (EP), (PERO), (PIIA) and the *completeness postulate* (CP) ((CP) For all $A\varepsilon\mathfrak{A}$ we have $S_Z(A)\neq\emptyset$). Then for each $A\varepsilon\mathfrak{A}$ there is a complete preference relation $\preccurlyeq_A$ on A with

(i) $a\varepsilon S_Z(A) \Rightarrow b \preccurlyeq_A a$ for all $b\varepsilon A$

(ii) The relations $\preccurlyeq_A$ induce a complete preference relation on $\bigcup_{A\varepsilon\mathfrak{A}} \text{Im } Z^A$ monotone with respect to the pre-ordering of R^n such that for all $A\varepsilon\mathfrak{A}$ the set $Z^A(S_Z(A))$ equals the set of greatest elements of Im Z^A with respect to this preference relation.

Proof: In the following we suppress the index A without any loss of generality.

a) Definition of the relation: $a,b \varepsilon A$

$$a \prec b \iff a \notin S_Z(\{a,b\}) \text{ and } b \varepsilon S_Z(\{a,b\})$$
$$a \sim b \iff a \varepsilon S_Z(\{a,b\}) \text{ and } b \varepsilon S_Z(\{a,b\})$$
$$a \preccurlyeq b \iff a \prec b \text{ or } a \sim b.$$

The case where neither a nor b belongs to $S_Z(\{a,b\})$ cannot happen according to (CP) and to the definition of a solution.

b) The completeness of the relation is implied by a).

c) Reflexivity is trivially fulfilled.

d) Transitivity: Let be $a \preccurlyeq b$ and $b \preccurlyeq c$. Hence, by definition, we have $b \varepsilon S_Z(\{a,b\})$ and $c \varepsilon S_Z(\{b,c\})$. From (PIIA) follows

$$S_Z(\{a,b,c\}) \cap \{b,c\} = \begin{cases} \emptyset & \text{case (1)} \\ S_Z(\{b,c\}) & \text{case (2)} \end{cases}$$

(1) => $S_Z(\{a,b,c\}) = \{a\}$ and by (PIIA) $S_Z(\{a,b,c\}) \cap \{a,b\} = \{a\} = S_Z(\{a,b\})$ and consequently $b \notin S_Z(\{a,b\})$ contrary to the assumption. Hence (2) must hold.

(2) => $c \varepsilon S_Z(\{b,c\}) = \{b,c\} \cap S_Z(\{a,b,c\})$, hence $c \varepsilon S_Z(\{a,b,c\})$ and consequently by (PIIA) we have $c \varepsilon S_Z(\{a,b,c\}) \cap \{a,c\} = S_Z(\{a,b\})$ and therefore $a \preccurlyeq c$.

e) Proof of (i): let be $a \varepsilon S_Z(A)$ and $b \varepsilon A$; $S_Z(A) \cap \{a,b\}$ contains a and thus by (PIIA) we have $a \varepsilon S_Z(A) \cap \{a,b\} = S_Z(\{a,b\})$ and there fore $b \preccurlyeq a$.

f) Up until now no use has been made of the conditions (EP) and (PERO); the remaining assertion (ii) is immediately proved by using these conditions. ¶

Solutions meeting the postulates (EP), (PERO), (PIIA) and (CP) with respect to the vector maximum problem ($\mathfrak{A}$,Z) are refered to as *rational with respect to* ($\mathfrak{A}$,Z). By theorem 2.1. we have shown a close relationship between rational solutions of vector maximum problems and classical economic rationality of complete preference relations. At the same time this theorem justifies the approach carried out somewhat later when we will develop a general solution approach immediately based on the preference relation in R^n guaranteed by theorem 2.1..

Remark: In case of uncertainty a solution concept may be formulated in full analogy. In that case the postulate of efficiency corresponds to the dominance principle of decision theory. An analogon to theorem 2.1. can be proved leading to a rationality weaker than the one required by the Bernoulli principle. A further discussion of this concept will be suppressed since the resulting approach is not yet operational.[1)]

1) Cf. Fandel and Wilhelm [1974].

Definition 2.5. The mapping $\eta : \wp(R^n) \to \wp(R^n)$ defined by

$$\eta(X) = \{y \varepsilon R^n \mid \bigvee_{x \varepsilon X} y \leq x\} \text{ for all } X \subset R^n$$

will be called the *refillment operation.*[1] The mapping $\text{eff} : \wp(R^n) \to \wp(R^n)$ defined by

$$\text{eff}(X) = \{y \varepsilon X \mid x \varepsilon X \text{ and } x \geq y \Rightarrow x = y\}$$

will be called the *efficiency operation.* eff(X) will be referred to as efficient *boundary* of X.

Lemma 2.1. $\text{eff} \cdot \eta = \text{eff}$.

Proof: Suppose $X \subset R^n$ and $x \varepsilon \text{eff}(X) \subset X \subset \eta(X)$. Let be $y \varepsilon \eta(X)$ and $x < y$; there is a $z \varepsilon X$ with $y \leq z$, thus $x < z$ contrary to the assumption. Hence $x \varepsilon \text{eff}(\eta(X))$. Conversely, let be $x \varepsilon \text{eff}(\eta(X)) \subset \eta(X)$ and $y \varepsilon X$ with $x \leq y$. By $X \subset \eta(X)$ we have $y \varepsilon \eta(X)$ and thus $y = x$ and, therefore, $x \varepsilon X$ and $x \varepsilon \text{eff}(X)$. ¶

Remark: $\eta \cdot \text{eff} = \eta$ fails in general. As a counterexample take a non-empty open set X in R^n then the efficient boundary of X is empty but the refillment of X is not. For a weaker statement we refer to the corollary of lemma 2.3 .

Throughout almost all the literature treating the vector maximum problem is based on the following condition which restricts the feasibility of vector maximum problems: given the vector maximum problem $(\mathfrak{A}, Z)$ we require each $A \varepsilon \mathfrak{A}$ to have at least one representation of the objective-system such that Im Z^A meets the following condition (CCB):

A subset Y of R^n is said to meet (CCB) iff

(C)¹ Y is *convex from below,* i.e. $\eta(Y)$ is convex.

(C)² Y is *closed from below,* i.e. $\eta(Y)$ is closed.

(B) Y is bounded from above and *weakly bounded from below,* i.e. there are elements $\bar{y}, \underline{y}$ R^n such that $\eta(Y) \subset \eta(\{\bar{y}\})$ and $\underline{y} \leq x$ for all $x \varepsilon \text{eff}(Y)$.

Then the set $Y_{\underline{y}} = \{x \varepsilon \eta(Y) \mid \underline{y} \leq x\}$ is convex, compact and we have $\text{eff}(Y_{\underline{y}}) = \text{eff}(\eta(Y)) = \text{eff}(Y)$ by lemma 2.1..

To prevent any confusion arising, we frequently say that $(\mathfrak{A}, Z)$ itself meets the condition (CCB). The condition is essentially to guarantee

1) This operation is used in game theory constructing characteristic functions; Aumann [1967, p. 6].

the existence of solutions, i.e. the postulate (CP); with some further assumptions on S_Z the sets $Z^A(S_Z(A))$ become bounded and convex non-empty subsets of R^n.

2.4.2. *The Treatment of The Vector Maximum Problem in the Literature*[1)]

The solution concepts advanced by the literature will be dealt with along the following criteria:

(1) Solution concepts violating any of our postulates (EP), (PERO), (PIIA) or (CP) with respect to vector maximum problems meeting (CCB), i.e. non-rational solutions;

(2) solution concepts fulfilling all these postulates: rational solutions; two subcriteria are applied:

   (a) solution concepts defined with respect to problems which do not meet the additional requirement of theorem 2.1.;

   (b) solution concepts immediately based on the preference relation guaranteed by theorem 2.1 .

Solution concepts of type (1) are to be repelled because they have unreasonable properties; the methods of this type are goal-programming[2)], methods of minimizing a geometrical distance function[3)], the chance-constrained programming[4)], the approaches of Aubin and Näslund [1972], of Benayoun et al. [1971] and of Belenson and Kapur [1973]. Until now one single approach of type (2a) has been suggested by Fandel [1972]: this approach is rational (from the formal standpoint) without leading to a preference relation.[5)]

---

1) For an extensive view of the literature on multi-objective decision making we may refer to Johnson [1968]. Some recent approaches are to be found in Fandel [1972, pp. 17-50]; a detailed discussion of the lately published concepts by using the instruments developed in 2.4.1. is presented by Fandel and Wilhelm [1974].

2) This notion is due to Charnes and Cooper [1967, Vol. I p. 215]. Further approaches to the goal-programming have been suggested by Balderstone [1960] and Ijiri [1965, pp. 34-36 and 43-45]. Similar but a little different is the approach of Sauermann and Selten [1962]; they do not minimize a geometrical distance function, but their solution especially violates (PIIA); the solution technique is somewhat related to those used to solve problems of type (2b): only local information at single points is required - naturally, not information provided by a preference relation.

3) Dinkelbach [1971, pp. 7-9]

4) Näslund [1967]

5) Cf. Fandel and Wilhelm [1974]

In the literature, the vector maximum problem is above all treated as a problem of computing the sets of solutions, i.e. as a problem of algorithms and rather secondarily as a problem of finding adequate solution concepts. Along this line two main ways to attack the problem are to be distinguished: either algorithms are searched for in order to compute solutions of type (2b), or it is questioned which algorithm computes solutions in the "best" way, yet, which is not precise, generating the solution concept by constructing the algorithm first. In both cases the following question is crucial: what information is required to be available? As a well-known fact the concepts "continous preference relation" and "continuous utility function" are mathematically equivalent, but they considerably differ with respect to the informational content. Knowing the preference relation might mean, e.g. that the decision maker is always able to choose amongst two given alternatives in a consistent way, but knowing the utility function requires the decision maker to evaluate every given alternative numerically. It turns out that the information required by the algorithms is the crucial point on which constructing methods compute solutions. We will present two examples suggested by the literature.

The solution concept and the corresponding algorithm suggested by Fandel (1972) are based on the assumption that, with respect to each given vector of objective-achievements, the decision maker has a certain idea for which of the objectives - assuming there are conflicting objectives - he does not accept any loss of achievement. These ideas are used to construct additional restrictions at each step of an iterative solution procedure such that the space of potential solutions shrinks into a point.

In order to compute solutions in the sense of (2b) Geoffrion [1965] and [1970] has reformulated the well-known gradient method of non-linear programming; the algorithm requires the decision maker to provide the marginal rates of substitution at each presented vector of objective-achievements and thus to provide the gradient of the utility function at this point (which indicates the direction of the "steepest" ascent of utility); by using this direction the algorithm computes a "best" feasible direction in order to determine the next starting point; as a second information the decision maker has to appoint how far it is to stride ahead in the indicated direction.

Searching for procedures possibly helpful for solving the problem under uncertainty, only methods based on solutions in the sense of (2b) may be accepted since our treatment of the problem on principle involves a monotone, complete preference relation. Supposing such a preference relation on $\bigcup_{A \in \mathfrak{A}} \text{Im } Z^A$ to be given, the problem of computing solutions defined by this relation may on principle be attacked in two ways: in the first place one can try to get along with local properties of the relation which can at all be experimentally determined; in the second place, under additional assumptions, one can fix the global structure by local properties and can, then, handle the problem as an analytical one.

Suggestions for the first method have been made by Marglin [1966] and by Geoffrion [1965] and [1970].[1] The information they require is hardly available, as far as we can see.[2]

The second approach has been dealt with by several authors.[3] This approach has the disadvantage that only special types of preference relations can be taken in; as soon as the decision maker's preferences are not of this type the problem cannot be solved by this method. But since until now there were no satisfying approaches enabling the decision maker to discover the whole preference structure by experimenting within a multi-dimensional choice-making situation, one can only analyze certain well-known structures and fit them to the decision maker's individual properties as well as possible.[4][5] The knowledge of the preference structure on the space of feasible vectors of objective-achievements of a decision situation under certainty is as we have seen the qualification of the solution of the problem under uncertainty, as now we have to determine the preference relation Q (of a superobjective) on the set of consequences $\text{Im } Z^\Gamma$ (for this objective). $\text{Im } Z^\Gamma$ or $Z^\Gamma(\bigcup_{a \in A} \{a\} \times Y_{\Gamma,a})$ respectively have here to be dealt with in place of $\text{Im } Z^A$.

But in a case to be analyzed later in more detail (case of indifference against objective-risks) a formal identity of the problem of multi-criteria decision making under uncertainty and the vector maximum problem

1) Under the assumption of linear utility functions and linear restrictions the problem has recently been treated by Zionts and Wallenius [1974].
2) Cf. Fandel and Wilhelm [1974].
3) A survey is given by Johnson [1968, pp. 421-434].
4) In order to determine the indifference hypersurfaces the method of "revealed preferences" presupposes that the decision maker can solve some non-trivial decision problems of the mentioned kind, without being able to describe the pattern of his indifference hypersurfaces.
5) Analogously, the portfolio selection theory replaces the general concept of a v.Neumann-Morgenstern utility function by a quadratic model where the coefficient of the quadratic term remains to fit the decision maker's specific risk-aversion.

takes place; hence, in this special case all methods may be applied which are suitable to solve the vector maximum problem - provided, naturally, the compatibility with preference relations is met. We will, now, suggest a new approach starting from fairly slight assumptions on the preference structure and requiring the decision maker to provide easily attainable information.

2.4.3. *The 2-dimensional Vector Maximum Problem*

Both goals indicated above: solving the n-dimensional vector maximum problem and fitting certain types of parameterized preference structures to the decision maker's specific properties may be carried out with the aid of the algorithm for the 2-dimensional vector maximum problem here to be presented. As prerequisites we need some results in a more special form well-known with the micro-economic theory.

2.4.3.1. *Abstract Treatment of the Problem*

Lemma 2.2. Let $Y \subset R^n$ be a closed set, bounded from above and weakly bounded from below (p. 62). Let $\preccurlyeq$ be a complete preference relation on R^n with

(i) $\preccurlyeq$ is continuous: $\{y \varepsilon R^n | y \preccurlyeq x\}$ and $\{y \varepsilon R^n | x \preccurlyeq y\}$ are closed sets for all $y \varepsilon R^n$.

(ii) $\preccurlyeq$ is monotone with respect to the preordering $\leq$ of R^n: $x < y \Rightarrow x \prec y$.

Then there are greatest elements of Y with respect to $\preccurlyeq$ and the set M of these elements is a subset of the efficient boundary eff(Y) of Y.

Proof: (1) If $x \varepsilon Y$ is a greatest element of Y with respect to $\preccurlyeq$, we have $x \varepsilon \mathrm{eff}(Y)$, for, suppose $y \varepsilon Y$ with $x < y$ then $x \prec y$ holds because of (ii) contrary to the assumption.

(2) Let $\underline{y}$ be such that $\underline{y} \leq \mathrm{eff}(Y)$. The set $\underline{Y} := \{x \varepsilon \eta(Y) | \underline{y} \leq x\}$ is compact by assumption. By theorem A of the appendix there is a greatest element z of $\underline{Y}$ with respect to $\preccurlyeq$. By (1) and by $\mathrm{eff}(Y) = \mathrm{eff}(\underline{Y})$ we have $z \varepsilon \mathrm{eff}(Y)$ in consequence of lemma 2.1.. A fortiori $x \preccurlyeq z$ holds for all $x \varepsilon \mathrm{eff}(Y)$; since for each $y \varepsilon Y$ there is a $x \varepsilon \mathrm{eff}(Y)$ (see the following lemma) with

$y \leq x$ and consequently, by (ii), with $y \preccurlyeq x$, we finally have $y \preccurlyeq z$ for all $y \varepsilon Y$; hence z is a greatest element of Y too. ¶

Lemma 2.3. Let $Y \subset R^n$ be a closed set bounded from above. Then to each $y \varepsilon Y$ there is a $x \varepsilon eff(Y)$ with $y \leq x$.

Proof: Let y be an arbitrary element of Y. The set $A:=\{z \varepsilon \eta(Y) | y \leq z\}$ is compact. Let $\bar{y} \varepsilon R^n$ be an upper bound of Y. Then the set $B:=\{\alpha y+(1-\alpha)\bar{y} | 0 \leq \alpha \leq 1\}$ is compact; hence $A \cap B$ is compact. The map $\alpha y+(1-\alpha)\bar{y} \to \alpha$ is continuous and has a minimum at $z \varepsilon B$. Naturally $z \varepsilon eff(Y)$ and $y \leq z$ hold. ¶

Corollary: Let Y be given as in lemma 2.3.. Then $\eta(eff(Y))=\eta(Y)$ holds.

The proof is obvious.

Theorem 2.2. Let be $Y \subset R^n$ and let Y meet the condition (CCB). Let $\preccurlyeq$ be a complete preference relation on R^n with (i) and (ii) of lemma 2.2 . Additionally let the following hold:

(iii) $\preccurlyeq$ is convex: $x \preccurlyeq y \Rightarrow x \preccurlyeq \alpha x + (1 - \alpha)y$ for all $\alpha \varepsilon <0,1>$

Then the set M of greatest elements of Y with respect to $\preccurlyeq$ is non-empty and convex.

If condition (C)[1] of (CCB) is replaced by

(SC) $\eta(Y)$ is convex and for all $x,y \varepsilon eff(Y)$ with $x \neq y$ and for all $\alpha \varepsilon >0,1<$ we have $\alpha x + (1 - \alpha)y \notin eff(Y)$

or if (iii) is replaced by

(iii') $\preccurlyeq$ is strictly convex: (iii) holds and $x \sim y$ implies $x \prec \alpha x + (1 - \alpha)y$ for all $\alpha \varepsilon >o,1<$,

then M consists of one and only one element.

Proof: (1) By lemma 2.2. M is non-empty and $M \subset eff(Y)$ holds. Let $x,y \varepsilon M$ be some elements with $x \neq y$. For all $\alpha \varepsilon <0,1>$ we have $z:=\alpha x + (1 - \alpha)y \in \varepsilon \eta(Y)$. By (iii) we have $x \preccurlyeq z$. By lemmata 2.1. and 2.3. there is a $w \in \varepsilon eff(\eta(Y)) = eff(Y)$ with $z \leq w$ and thus $z \preccurlyeq w$. Suppose $z < w$, then $z \prec w$ and therefore $x \prec w$ must hold contrary to the assumption. Hence $z = w$, $z \varepsilon Y$ and $z \varepsilon M$ hold from what we have the convexity of M.

(2a) Suppose (SC) to hold. Let be $x,y \varepsilon M$ and $x \neq y$; then by (1) convex combinations of x and y are in M and hence by lemma 2.2. efficient points contrary to (SC); hence x equals y.

(2b) Suppose (iii') to hold; let be $x,y \varepsilon M$ with $x \neq y$. Then convex combinations of x and y belong to M and are, by (iii'), strictly prefered to x which is impossible; hence x = y must hold. ¶

Replacing Y by the price simplex the derived results are in a more special form well-known facts in the micro-economic theory of consumer behavior.[1]

Lemma 2.4. Let $\langle a,b\rangle$ be a compact interval of R, $F : \langle a,b\rangle \rightarrow R$ a strictly monotone decreasing function which meets the following condition for all $x,y \varepsilon \langle a,b\rangle$ and all $\alpha \varepsilon \langle 0,1\rangle$:

(i) $F(\alpha x + (1 - \alpha)y) \geq \alpha F(x) + (1 - \alpha)F(y)$

Additionally, let F be continuous at b.
Then

$$Y := \bigcup_{x \varepsilon \langle a,b\rangle} \{x\} \times \langle F(b), F(x)\rangle$$

meets the condition (CCB). If, additionally, in (i) the symbol $\geq$ is replaced by >, condition (SC) of theorem 2.2 is fulfilled. In either case $eff(Y) = \{(x,F(x) \mid x \varepsilon \langle a,b\rangle\}$ holds.

The proof may be left to the reader.

Remark: Lemma 2.4 may be formulated in higher dimensions too, if the interval is replaced by a convex and compact set and F is assumed continuous at all boundary points. But we will make use of lemma 2.4 in the given formulation only.

2.4.3.2. *The Algorithm for the Case of Two Dimensions*[2]

Throughout this paragraph let F be a function meeting the conditions of lemma 2.4 and $\preccurlyeq$ be a continuous, convex and complete preference relation on R^2. We will make use of lemma 2.4 in the form of the strict inequality in (i) or assume (iii') instead of (iii) for the preference relation. Let Y be defined by lemma 2.4 ; then, by theorem 2.2 , there is one and only one greatest element $x^+ \varepsilon Y$ with respect to $\preccurlyeq$. We intend to find this point

a) by a countable number of comparisons with respect to $\preccurlyeq$, and

1) Cf. Debreu [1959]
2) For a similar algorithm concerning the chance-constrained programming problem we refer to Näslund [1967, pp. 34-37]. In order to provide the marginal rates of substitution for the algorithm of Geoffrion (p. 64) Dyer [1972, p. 206] and [1973] has suggested a similar approach.

b) by computing the value of F at a countable number of points;

c) the optimal point x^+ has to be approximated by this procedure at any desired (positive) degree of accuracy by means of a finite number of steps.

To this end we use the following simple argument:

(R1) $x=(x_1,F(x_1))$; $y=(y_1,F(y_1))$; if $x_1,y_1 \leq x_1^+$ and $x_1 < y_1$ hold, we have $x \prec y$.

For a proof we construct a convex combination of x and x^+ which is less than or equal to y; such a combination must exist because of condition (i) of lemma 2.4 . As $\preccurlyeq$ is convex x is not prefered to this combination and thus, by monotoneicy and transitivity to y. It is easily shown as a consequence of (SC) or (iii') that even $x \prec y$ holds.

In the same way the following is seen:

(R2) Let x and y be like in (R1); let $x_1,y_1 \geq x_1^+$: if $x_1 < y_1$ holds we have $y \prec x$.

Notation: Instead of $(x,F(x)) \preccurlyeq (y,F(y))$ we write $x \preccurlyeq y$.

Let $i \varepsilon N$ be fixed; $<a_i,b_i> \subset <a,b>$ and $x_1^+ \varepsilon <a_i,b_i>$. Let $x_i,y_i \varepsilon >a_i,b_i<$ with $x_i < y_i$. The following three cases may occur:

(1) $x_i \leq x_1^+$ and $y_i \leq x_1^+$ which implies $a_i \prec x_i \prec y_i$ by (R1)

(2) $x_1^+ \leq x_i$ and $x_1^+ \leq y_i$ which implies $b_i \prec y_i \prec x_i$ by (R2)

(3) $x_i \leq x_1^+$ and $x_1^+ \leq y_i$ which implies $a_i \prec x_i$ and $b_i \prec y_i$.

A) Suppose $a_i \sim b_i$ not to hold; then we have

(F1) $x_i \sim y_i$ implies $x_1^+ \varepsilon >x_i,y_i<$: set $(a_{i+1},b_{i+1}) = (x_i,y_i)$!

(F2) $a_i \prec x_i \prec y_i$ implies $x_1^+ \varepsilon <x_i,b_i>$: set $(a_{i+1},b_{i+1}) = (x_i,b_i)$!

(F3) $b_i \prec y_i \prec x_i$ implies $x_1^+ \varepsilon <a_i,y_i>$: set $(a_{i+1},b_{i+1}) = (a_i,x_i)$!

In either case set $x_{i+1} = 2/3\ a_{i+1} + 1/3\ b_{i+1}$ and $y_{i+1} = 1/3\ a_{i+1} + 2/3\ b_{i+1}$. No further cases may occur under the above-made assumption.

B) If $a_i \sim b_i$ holds we have to check an additional point z_i between x_i and y_i to come up with analogous conclusions which can be left to the reader.

From A) and B) a procedure is easily derived which concludes from simple preference statements the partial interval the optimal point must belong to and restricts the further consideration to this partial interval repeating this process until an interval of desired maximal length is arrived at.

By our preceding arguments we have shown that the optimal point belongs to the interval under consideration for each iteration. Trivially the process converges, i.e. the considered intervals shrink to the optimal point, since the interval under consideration may be reduced by at least a quarter of its length at each iteration. At each step the decision maker has to carry out at most ten comparisons of two efficient points at a time. Thus the indicated conditions a), b) and c) are met.

The problem of 2-objective decision making under certainty has consequently been solved under quite general assumptions; neither the whole preference structure nor the whole efficient boundary has had to be explicit. The developed algorithm will be used to determine special preference structures in R^n as a whole as well as to construct an algorithm for the general vector maximum problem. The technical problems of the rapidity of convergence and of intransitivities possibly caused by the decision maker will not be analysed, since we are more interested in the principal treatment of the problem.

2.4.4. *Linear Preference Structures and the Determination of the Weights*

In this section we will investigate the case of linear preferences. At this time we are not interested in the question under which circumstances linear preferences are a reasonable representation of the decision maker's "true" preferences.

By *linear preferences* on R^n we mean preference relations generated in the following way: there is a vector $a \in R^n$, the *weighting vector*, such that $x \preccurlyeq y \iff a.x \leq a.y$. In order to have $\preccurlyeq$ monotone we assume $a \gg 0$. Dealing with linear preferences involves two problems:

1. Which vector a is to be chosen in order to describe the decision maker's preference behavior correctly ?
2. Which are the consequences of the fact that the space of imaginable vectors of objective-achievements is unique only up to certain transformations associated to the system of objectives under consideration ?

Let us start with problem 2. Let $(\mathfrak{A},Z)$ be a pair standing for a vector maximum problem or for a problem of multi-criteria decision making under uncertainty (in the latter case Z is a complete system of objectives). To each (partial) objective i corresponds a set T_i of transformations in R. We confine ourselves to the cases where T_i are subgroups of $L_+(R)$, the group of all positive linear transformations of R; that means that the measurement of the partial objectives is unique up to the origin and the unity of the scale. In "higher" cases of linear transformations the linear model becomes uncomfortable. As long as no purely qualitative criteria are considered this limitation does not involve any severe consequences. Thus $\prod_i T_i$ is a group of linear transformations in R^n.

Let

$$M := \begin{cases} A & \text{be under certainty} \\ \bigcup_{a\in A} \{a\}\times Y_{\Gamma,a} & \text{be under uncertainty} \end{cases}$$

where $\Gamma=(A,\langle\Omega,\mathcal{A},\mu\rangle_A)\in\mathfrak{A}$. Then the system of objectives Z determines a function $Z^M : M \to R^n$, unique up to transformations in T, i.e. each function $t\cdot Z^M$ with $t\in T$ belongs to another representation of the same system of objectives; the "content" related to these functions is the same for each $t\in T$.

If it is assumed that the decision maker has a preference relation on R^n - interpreted as the set of imaginable vectors of objective-achievements with respect to the representation Z^M - this relation does not, as a rule, remain unchanged turning from Z^M to $t\cdot Z^M$. Therefore, the preference relation is heavily based on the considered representation; the relation on M is invariant only.

Let Z^M be an arbitrary representation of the system of objectives; let the corresponding relation $\preccurlyeq_{Z^M}$ in R^n be linear with the generating weighting vector a_{Z^M}.

Assertion: For each $t\in T$ the relation $\preccurlyeq_{t\cdot Z^M}$ is linear too.

Proof: $t\in T$ can be represented by a matrix in diagonal form t^1 and a vector $t^2\in R^n$: $t(x) = t^1x + t^2$. The following must hold: $x \preccurlyeq_{Z^M} y \iff t(x) \preccurlyeq_{t\cdot Z^M} t(y)$. Choosing $a_{t\cdot Z^M} := a_{Z^M}\cdot {t^1}^{-1}$ we have $a_{t\cdot Z^M}\cdot t(x)=a_{Z^M}\cdot x + c$ where c is a constant independent of x. Using this equation the assertion is easily proved. At the same time we have found the formula for transforming the weighting vector. ¶

Proposition 2.3. Let K^n be the unit-n-sphere: $K^n=\{x\in R^n \mid \|x\| \leq 1\}$, where

$\| \;\|$ stands for the euclidean Norm of R^n. Let be $a \varepsilon R^n$ with $a >> 0$. Let for x^+ the following hold:

$$a.x^+ = \max\{a.x \mid x \varepsilon K^n\}.$$

Then we have $x^+ = a/\| a \|$.

Proof: We form the Lagrangian $L = a.x + \lambda(x.x - 1)$. For the optimum x^+ we must have

$$0 = \frac{\partial L}{\partial x_i}(x^+) = a_i + 2\lambda x_i^+ \qquad (i=1,\ldots,n)$$

and

$$x^+.x^+ = 1.$$

Now the assertion is immediately seen. ¶

Since parameter vectors a and b generate the same linear preference structure if there is a positive number w with $a = wb$, the linear preference structure in R^n is uniquely determined with respect to the chosen representation by giving the optimal point of the sphere of unity.

Linear preferences on R^n meet the following condition:

Independency of the Number of Coordinates (INC)[1]

Let $\preccurlyeq$ be some complete preference relation on R^n; suppose $i < j$ and $i,j \varepsilon \{1,\ldots,n\}$; let be $x_m, y_m \varepsilon R$ where $m \varepsilon \{1,\ldots,n\}$ and $m \neq i,j$. We define the following relations on R^2:

$$z \preccurlyeq_x w \iff z_x \preccurlyeq w_x \quad \text{where}$$

$$pr_m z_x := \begin{cases} z_1 & \text{if } m=i \\ z_2 & \text{if } m=j \\ x_m & \text{elsewhere} \end{cases}$$

and analogously for y. $\preccurlyeq$ is said to meet (INC) if for all choices of i,j,x,y of the mentioned kind $\preccurlyeq_x$ equals $\preccurlyeq_y$.

Proposition 2.4. Linear preferences on R^n meet (INC).

Proof: Let $\preccurlyeq$ be linear with weighting vector a. Let i,j,x,y be given as in (INC); $z,w \varepsilon R^2$. Then we have $z \preccurlyeq_x w \Longleftrightarrow z_x \preccurlyeq w_x \Longleftrightarrow a.z_x \leq a.w_x$. $a.z_x = a_i z_1 + a_j z_2 + \sum_{m \neq i,j} a_m x_m$ and analogously for $a.w_x$, $a.z_y$ and $a.w_y$. Hence we have $a.z_x \leq a.w_x \Longleftrightarrow a_i z_1 + a_j z_2 \leq a_i w_1 + a_j w_2 \Longleftrightarrow a.z_y \leq a.w_y$ which completes the proof. ¶

Proposition 2.3. and proposition 2.4. enable us to determine the weigh-

---

1) This condition is equivalent to postulate E of Fleming [1952].

ting vector a of a linear preference structure with respect to a given representation of the system of objectives.

Given the representation Z^M the unit-n-sphere K^n will be interpreted as a set of feasible Z^M-vectors. For each pair $(i,i+1)$ we consider R^2 a subset of R^n determined by the condition $x_j=0$ for all $j\neq i,i+1$. The linear preference structure on R^2 dtermined by (a_i,a_{i+1}) equals the trace of the relation $\preccurlyeq$ generated by $(a_1,\ldots,a_n)$ on R^n since $\preccurlyeq$ meets (INC) by proposition 2.4 . Applying proposition 2.3 for n=2 we have for the optimal point (z_i,z_{i+1}) of K^2

$$\frac{z_i}{z_{i+1}} = \frac{a_i}{a_{i+1}} .$$

The point (z_i,z_{i+1}) may be determined by the algorithm set forth in 2.4.3.2. at the same time leading to the quotient a_i/a_{i+1}; proceeding in this way for all $i=1,\ldots,n-1$ and taking $a_1=1$ we are able to compute the weighting vector a which is unique up to positive constant multipliers.

Remark: Since the algorithm for the determination of the optimal point in the case of two dimensions comes up with this point generally up to a certain degree of accuracy only, it is necessary to state that the preferences depend in some sense continuously on the weighting vector. Furthermore, the vector a depends on the optimal point in a continuous way too, such that small errors made in determining the optimal point do not involve errors as great as you might think with respect to the representation of the preference structure.

As far as we see our approach enables us to determine the weights of a linear preference structure in a theoretically consistent manner. Another approach due to Churchman and Ackoff [1954] and [1959, p. 138] makes additional assumptions on the decision maker's ability in not being able to derive the weights conclusively. We will not deal with this anymore.[1)]

1) Another approach is to be found in Eckenrode [1965, p.180]. From the view of social preference functions Theil [1963] has suggested another method. Lately published approaches are due to Nievergelt [1971] and Zionts and Wallenius [1974].

2.4.5. *Hyperbolic Preferences and the Determination of the Exponential Weights*

By *hyperbolic preferences* on R^n we mean a preference relation on R^n_+ generated by the following condition: there is a vector $a>>0$, the *vector of exponential weights*, such that for all $x,y \in R^n_+$ we have

$$x \preccurlyeq y \iff \prod_{i=1}^{n} x_i^{a_i} \leq \prod_{i=1}^{n} y_i^{a_i}$$

Since $a>>0$ the preferences are monotone; if $a_i < 1$ for some i we choose the positive root if there exists more than one.

Here the same problems arise as in the preceding section:

1. Which is the parameter vector a related to the decision maker ?
2. How is it dealt with that the space of imaginable vectors of objective-achievements is determined up to certain transformations only which are related to the system of objectives ?

We will argue as in the preceding section: let T be the group of linear transformations associated to the system Z. In case of hyperbolic preferences we admit the subgroups of the group of dilatations only: T is a group of matrices in diagonal form with only positive elements. Let Z^M be defined as above; let $\preccurlyeq_{Z^M}$ be hyperbolic with weights a_{Z^M}.

Assertion: For each $t \in T$ the relation $\preccurlyeq_{t \cdot Z^M}$ is hyperbolic too.

Proof: For $t \in T$ we have $x \preccurlyeq_{Z^M} y \iff t(x) \preccurlyeq_{t \cdot Z^M} t(y)$; the symbol t is used for the diagonal matrix as well as for the vector of diagonal elements of this matrix. We fix $a_{t \cdot Z^M} = a_{Z^M} = a$ and denote by u the following function:

$$u(x) = \prod_{i=1}^{n} x_i^{a_i}.$$

Now it is easily seen that $u(t(x)) = u(t.x) = u(t).u(x)$ holds. The rest of the proof may be left to the reader. ¶

The function u is often called the *Cobb-Douglas utility function*. At the same time we have seen that turning frome one representation of the system to another leaves the vector of exponential weights unchanged.

Proposition 2.5. Let c be a real positive number and H^c_n the hyperplane in R^n consisting of the points whose coordinates sum up to c. Let u_a be the Cobb-Douglas utility function above-defined. Let x^+ be the (optimal) point of H_n^c maximising the utility on H_n^c. Then we have for all $i,j=1,\ldots,n$

$$\frac{x_i^+}{x_j^+} = \frac{a_i}{a_j} .$$

The proof is (again) a simple application of the Lagrange-multiplier theorem.

It is a remarkable fact that $u_a(x)^k = u_{ka}(x)$ holds for all $k > 0$; hence a and ka determine the same hyperbolic structure, the parameter a is unique up to positive multipliers only. We will follow this up later.

Proposition 2.6. Hyperbolic preferences meet (INC).

Proof: Let $\preccurlyeq$ be hyperbolic with weights a. Let i,j,x,y be given as in (INC). For $z,w \varepsilon R^2$ we have:

$$z \preccurlyeq_x w \iff z_x \preccurlyeq w_x \iff u_a(z_x) \leq u_a(w_x).$$

From

$$u_a(w_x) = \prod_{\substack{m=1 \\ m \neq i,j}}^{n} x_m^{a_m} . w_1^{a_i} w_2^{a_j}$$

the desired result is easily derived. ¶

In order to determine the exponential weights $a_1,\ldots,a_n$ we proceed as before in the case of linear preferences: we choose a $z \varepsilon H_n^c$ with $z \gg 0$; then the set $H_{ij}^c := \{x \varepsilon H_n^c | x_m = z_m$ for all $m \neq i,j\}$ is a straight line through z. The hyperbolic preferences generated by (a_i,a_j) on H_{ij}^c equals the trace of the relation generated by a on R_+^n (proposition 2.6.). H_{ij}^c may be understood as a hyperplane of R^2; using proposition 2.5 we find for the related optimal point (h_i,h_j):

$$\frac{h_i}{h_j} = \frac{a_i}{a_j} .$$

By a process of determining the optimal points of n-1 straight lines of R^n we finally achieve the vector a (up to a positive constant multiplier) as in the case of linear preferences.

Remark 1: The linear as well as the hyperbolic case meets the following conditions:

(1) complete, continuous preferences
(2) monotone preferences
(3) (INC)

These conditions are identical with the postulates (A) trough (E) used by Fleming [1952, p. 310] in dealing with a social welfare function. Fleming has shown that the (ordinal) utility function guaranteed by condition (1) may be written as a weighted sum of the (partial) objective functions, suitable monotone transformations of the partial objective functions provided, i.e. given that a suitable representation of the system of objectives is selected. In the light of this fact the connection of the depicted cases may be seen as follows:

$$\log u_a(x) = \sum_{i=1}^{n} a_i \log x_i .$$

The log-function is strictly increasing and continuous with respect to R_+^n from what u_a and $\log u_a$ represent the same preference structure in ordinal terms. Within our cardinal context (the feasible transformations of the objective functions are supposed to belong to $L_+(R)$) we see that both cases really differ. As an interesting fact we remark that the transformation group of dilatations and the full group $L_+(R)$ correspond exactly by the property of the log-function as an isomorphism of the multiplicative group R_+ onto the additive group R.

Remark 2: The definition of linear and hyperbolic preferences as well as the determination of the respective parameters has been based on the indifference hypersurfaces only; hence the related utility functions are ordinal, for monotone continuous transformations leave the form of the indifference hypersurfaces invariant. In order to treat the uncertainty problem we need cardinal utility functions; hence it remains to select the correct feasible transformation transposing the functions a.x and $u_a(x)$ into the appropriate v.Neumann-Morgenstern utility function, i.e. into the function u whose integral represents the preference relation on the set of probability measures on M. These problems will be dealt with later.

2.4.6. *Generalization*

In this section we will sketch out how the indicated approaches may be generalized. As remark 1 of the preceding section states, every preference relation fulfilling the indicated conditions may be described as follows: if u is a continuous utility function representing the preference relation under consideration, then there are strictly increasing continuous functions $g, f_1, \ldots, f_n : D \to R$ $(D \subset R)$ and real numbers $a_1, \ldots$ $\ldots, a_n$ such that we have for all x

$$g \cdot u(x) = \sum_{i=1}^{n} a_i f_i(x_i) .$$

Proposition 2.7. Let the preference relation generated by

$$u(x) = \sum_{i=1}^{n} a_i f_i(x_i)$$

be convex and let u be differentiable. Let $S^n := \{x \in R^n \mid \|x\| = 1\}$ be the surface of the unit-n-sphere. Let x^+ be the point where utility is maximized on S^n. Then we have for all $i, j = 1, \ldots, n$:

$$\frac{a_i}{a_j} = \frac{x_i^+}{x_j^+} \frac{\frac{df_j}{dx_j}(x_j^+)}{\frac{df_i}{dx_i}(x_i^+)} .$$

Proof: x^+ is uniquely determined by theorem 2.2 . Forming the Lagrangian we have the assertion as an immediate consequence of the necessary conditions for x^+. ¶

In order to determine the parameter vector a we form $S_{ij} = \{x \in S^n \mid x_m = 0, m \neq i,j\}$ and denote by $\bar{x}$ the point maximizing utility on S_{ij}. As an analogon to proposition 2.7 we have the quotient a_i/a_j given by an expression which comprises $\bar{x}_i, \bar{x}_j$ and the derivatives of f_i and f_j at $\bar{x}_i$ and $\bar{x}_j$ respectively. Hence, as a 2-dimensional problem $\bar{x}$ may be determined by the algorithm set forth in 2.4.3.2. iteratively computing the quotients a_i/a_{i+1} finally coming up with the vector a up to a positive constant multiplier.

At last we analyse the consequences of replacing one representation of the system of objectives by another, i.e. a transition from x_i to $h_i(x_i)$ with $h_i \in T_i$. Obviously, such a replacement involves a transition from f_i

to $f_i \cdot h_i^{-1}$. Accordingly, in general we do not have a transformational formula with respect to the parameter vector only whereas the functions f_i remain unchanged.

Result: All complete, monotone and continuous preferences meeting (INC) can be fitted to the decision maker with the aid of the 2-dimensional algorithm convexity and differentiability provided. But here again we have to refer to remark 2 of the preceding section: the problem of the correct utility function for an application of the expected-utility theorem remains to be solved in addition to the fitting procedure above desribed, for the transformation g might fail to be linear.

2.4.7. Sketch of an Algorithm to the n-dimensional Vector Maximum Problem

In this section we will give an application of the 2-dimensional algorithm developed in 2.4.3.2. to the general case of n dimensions. In so doing we will not prove all assertions in detail, but we will argue more intuitively in order to elucidate the basic ideas.

We consider again a set $Y \subset R^n$ representing the set of considered vectors of objective-achievements. Additionally we assume the decision maker to have strictly convex, monotone and differentiable preferences in R^n, i.e. preferences which can be represented by a monotone, concave, differentiable utility function. Let Y be a closed set bounded from above, weakly bounded from below, and let $\eta(Y)$ be convex. Then the efficient boundary of Y is one to one to a subset $\bar{Y}$ of R^{n-1} by the following projection map:

$$p(x_1,\ldots,x_n) = (x_2,\ldots,x_n).$$

Then we have

$$x \varepsilon \mathrm{eff}(Y) \iff (x_2,\ldots,x_n) \varepsilon \bar{Y} \text{ and } x_1 = \max\{y \varepsilon R \mid (y,x_2,\ldots,x_n) \varepsilon Y\}.$$

Naturally $\bar{Y}$ is compact in R^{n-1}; we assume $\bar{Y}$ convex and to have inner points in R^{n-1}. The preference relation $\preccurlyeq$ induces a preference relation on $\bar{Y}$ in a natural way (also denoted by $\preccurlyeq$) since p is a bijection. This relation is strictly convex on $\bar{Y}$ too because of the convexity of $\bar{Y}$. Additionally we suppose that eff(Y) is a smooth hypersurface in R^n, then $\preccurlyeq$ is differentiable on $\bar{Y}$.[1)]

1) eff(Y) is always differentiable up to a set of Lebesgue measure zero: Karlin [1962, Vol. I, p. 405].

The Algorithm

(i) Choose some straight line in R^{n-1} containing an inner point of $\bar{Y}$. For sake of simplicity a straight line parallel to the first coordinate axis is chosen. Because of the convexity of $\bar{Y}$ and of $\preccurlyeq$ we can determine the greatest element (with respect to $\preccurlyeq$) of the intersection of this line with $\bar{Y}$ by the 2-dimensional algorithm. We assume this point to be an inner point of $\bar{Y}$ and denote it by x^1.

(ii) Form a second line through x^1 parallel to the second coordinate axis and determine the related greatest element x^2.

(iii) Continue in this way up to the point x^{n-1} constructing in the i-th step a straight line through x^{i-1} parallel to the i-th coordinate axis. In the n-th step construct a straight line through x^{n-1} parallel to the first coordinate axis and continue in the indicated way.

This algorithm generates a sequence of points increasing with respect to the preference relation concerned. Since $\bar{Y}$ is compact and $\preccurlyeq$ is continuous there is a limit of this sequence, this limit point must be optimal with respect to $\preccurlyeq$: virtually, this is a consequence of the differentiability of $\preccurlyeq$.

If a point arising within the sequence is a boundary point of $\bar{Y}$, we must try to reach inner points again by "small" shiftings; for reasons of continuity this will on principle be possible, though it may, however, produce practical difficulties.

As soon as n-1 consecutive points coincide the optimal point is attained.

In the following charts the process is graphically illustrated for n=3.

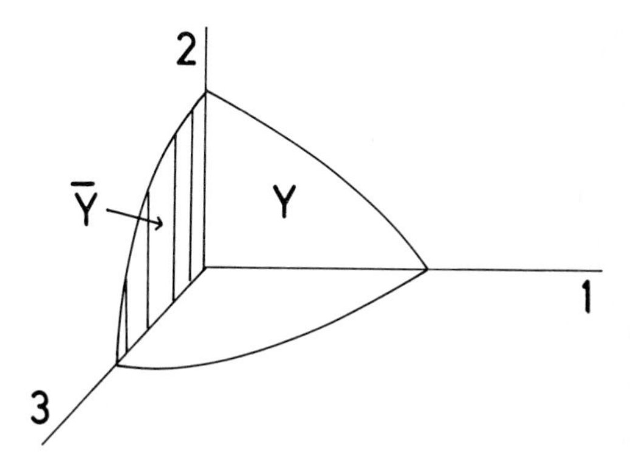

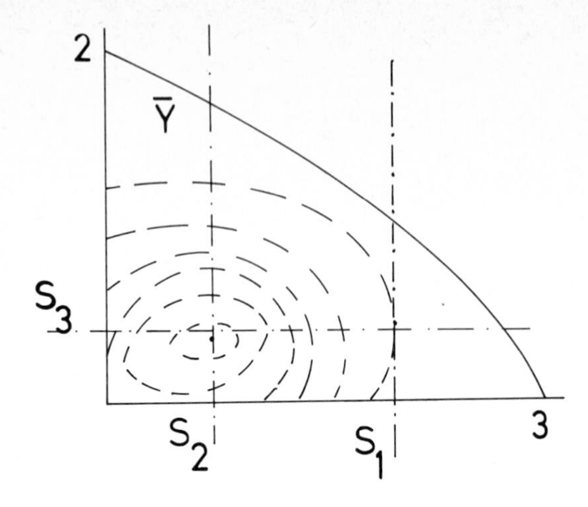

The rigorous proof of the optimality of the limit point is based on the fact that strictly convex sets with a differentiable boundary have one and only one supporting hyperplane through each boundary point which is at the same time tangent to this point. Given the optimal point of a straight line this line is tangent to some indifference hypersurface through this point. An indifference hypersurface cannot have more than n-2 linearly independent tangents to the same point, unless it shrinks to that point. Neither do we want to carry this out in more detail nor are we concerned with problems of boundary points arising within the iterative process. Assuming $\bar{Y}$ convex and eff (Y) differentiable may probably be weakened, at least when the problems are to solve where the efficient boundary is piecewise differentiable.

The question on the influence of the fact that the optimal point at each 2-dimensional step is only approximated (only a finite number of iterations is possible) will not be dealt with in more detail either; in carefully applying the algorithm this question will not be of great importance because of the continuity assumptions. Those more technical problems would have to be analysed for a successful application of the algorithm on real life decision problems; we were more concerned to show that the vector maximum problem seems solvable within the concept of preference relations under fairly weak conditions. The problem plays a secondary role for our purposes, since in special cases only, the problem of multi-objective decision making under uncertainty is formally identical with the vector maximum problem.

3. *Solution Approaches to the Problem of Multi-Objective Decision Making under Uncertainty*

We now give a short restatement of the problem. $\mathfrak{A}$ is a class of decision situations under uncertainty and Z a complete system of n partial objectives for $\mathfrak{A}$. By Z we also denote a representation of the n-dimensional criterion function of the system. Z is defined as a real vector function on the sets $\bigcup_{a \in A} \{a\} \times Y_{\Gamma,a}$ where $\Gamma = (A, <\Omega, \mathcal{A}, \mu>_A)$. The set $\prod_i T_i$ denoted by T is the set of monotone transformations in R^n related to the system of objectives. Every representation of the criterion function of Z may be described by $Z' = t \cdot Z$ for an appropriate $t \in T$. For $\Gamma \in \mathfrak{A}$ we denote by Z^Γ the restriction of Z to the situation Γ. Then for each $a \in A$ the function $Z^\Gamma(a,.) : Y_{\Gamma,a} \rightarrow R^n$ is measurable and μ_a induces on R^n a pro-

$$\mu_a^Z(B) := \frac{\mu_a(Z^\Gamma(a,.)^{-1}(B))}{\mu_a(Y_{\Gamma,a})},$$

where $B \in \mathcal{B}$ and $\mathcal{B}$ is some σ-algebra of R^n completely contained in the Borel-σ-algebra.

3.1. *The Linear Model*

As we have previously said we consider the closed convex hull of Im Z (more precisely of the intersection of Im Z and R^n; in the following we consider Z to be a real vector function in accordance with the remarks made in section 2.3) to be the set of consequences of a superobjective and denote it by I^n. Then the superobjective is given by a complete monotone preference relation Q on I^n and a utility function $u : I^n \rightarrow R$. But the utility function is assumed not to be fully explicit. Consequently we investigate the related preference relation on the set of probability measures on I^n. We assume the axioms of the expected-utility theorem 1.3 and of theorem 1.4 to hold. Postulating one additional condition determines the utility function guaranteed by theorems 1.3 and 1.4 .

Theorem 3.1. (due to Harsanyi) Let $\preccurlyeq$ be a continuous, monotone complete preference relation on I^n. Let I^n have inner points with respect to R^n. Let M be the set of probability measures on I^n with respect to the interval-σ-algebra of $\preccurlyeq$. Let the conditions (B1) through (B5) hold for $\preccurlyeq$; let the mapping $i : I^n \to M$ be defined as in lemma 1.4 . Suppose for all $x,y\varepsilon I^n$ and $\alpha\varepsilon\langle 0,1\rangle$:

(I) $\alpha i_x + (1 - \alpha)i_y \sim i_{\alpha x + (1 - \alpha)y}$.

Then the preference relation $\preccurlyeq$ is the trace of a linear preference relation on R^n: there is a vector $a>>0$ and a constant $c\varepsilon R$ such that the utility function $u : I^n \to R$ can be written as $u(x) = a.x + c$.

Proof:[1] Without any loss of generality we may suppose $0\varepsilon I^n$ and 0 to be an inner point of I^n, for a linear preference ordering remains linear with respect to a positive linear transformation (2.4.4.). Let u be the utility function guaranteed by theorems 1.3 and 1.4 .

Assertion: If $x\varepsilon I^n$ and $\alpha\varepsilon R$ are such that $x\varepsilon I^n$ we have $u(\alpha x) - u(0) = \alpha(u(x) - u(0))$.

The proof is given in several steps:

1) $\alpha\varepsilon\langle 0,1\rangle$: $u(i_{\alpha x})=u(\alpha i_x + (1 - \alpha)i_0)=\alpha u(i_x) + (1 - \alpha)u(i_0)$. Hence $u(\alpha x)=\alpha u(x) + (1 - \alpha)u(0)$ and $u(\alpha x)=\alpha(u(x) - u(0)) + u(0)$.

2) $\alpha=-1$. Let $\beta\varepsilon\rangle 0,1/2\langle$, then we have $\beta i_x + (1 - \beta)i_{-x} \sim i_{\beta x - (1 - \beta)x} = i_{x(2\beta - 1)}$. Hence $\beta u(x) + (1 - \beta)u(-x)=u(x(2\beta - 1))$. By 1) we have $u((1 - 2\beta)(-x))=(1 - 2\beta)(u(-x) - u(0)) + u(0)=(1 - 2\beta)u(-x) + 2\beta u(0)$. Hence $\beta u(x) + (1 - \beta)u(-x)=(1 - 2\beta)u(-x) + 2\beta u(0)$ and $\beta u(x)=-\beta u(-x) + 2\beta u(0)$ and finally $u(x) - u(0)=-u(-x) - u(0))$ hold.

3) The case $\alpha < 0$ and $\alpha \geq -1$ is proved by combining 1) and 2).

4) $\alpha > 1$. $y:=\alpha x$. We have $1/\alpha < 1$ and $y\varepsilon I^n$ by assumption. Furthermore we have $i_x \sim i_{\alpha^{-1}y}$ and thus $u(x)=\alpha^{-1}u(y) + (1 - \alpha^{-1})u(0)$, $u(x) - u(0) = \alpha^{-1}(u(y) - u(0))$ and $\alpha(u(x) - u(0))=u(\alpha x) - u(0)$.

5) The case $\alpha < -1$ is proved by combining 2) and 4).

---

1) For the following arguments we refer to Harsanyi [1955, p. 312]. The formulation of the theorem given here slightly differs from Harsanyi's exposition; Harsanyi deals with the problem of a social welfare function. Accordingly, he interprets the coordinates of R^n as individual utilities.
Rothenberg [1961, chap. 10] argues that the axioms of a v.Neumann-Morgenstern utility function are not as plausible for a social preference system as for individual preferences.

For the rest of the proof we assume $u(0) = 0$ without any loss of generality. Let $\{e^i\}_{i=1,\dots,n}$ be an orthogonal basis of R^n such that for all i the vectors $\pm e^i$ belong to I^n. Such a basis exists since 0 has been assumed to be an inner point of I^n. Let us define $a_i = u(e^i)$. For each $x \epsilon I^n$ we have a unique representation $x = \sum_i x_i e^i$. We define $z := \sum_i |x_i|$, then we have for all i: $z^{-1} x_i e^i \epsilon I^n$ and thus the preceding arguments show $u(z^{-1}x) = \sum_i u(z^{-1} x_i e^i)$ and therefore $z^{-1} u(x) = z^{-1} \sum_i x_i u(e^i)$ and finally:
$u(x) = \sum_i a_i x_i$. This completes the proof. ¶

Theorem 3.1 provides exactly what we want: the probability measures generated by Z(b,.) on R^n for each $b \epsilon A$ may be understood as elements of M, if the interval-σ-algebra of $\preccurlyeq$ is taken for B. Under the condition of theorem 3.1 there exists a positive vector $a >> 0$ such that the decision is made in accordance to

$$U_a(b) := \int_{Y_{\Gamma,b}} a.Z^{\Gamma}(b,.) \, d\mu_b$$

$$= a. \int_{Y_{\Gamma,b}} Z^{\Gamma}(b,.) \, d\mu_b \qquad (b \epsilon A).$$

$U_a(b)$ is unique up to positive linear transformations.

Particularly, the decision is made by using the expected values of the partial objectives only: defining

$$z^{\Gamma}(b) := \int_{Y_{\Gamma,b}} Z^{\Gamma}(b,.) \, d\mu_b$$

we have $U_a(b) = a.z^{\Gamma}(b) + c$ where c is some constant in R.

Here we must remember our arguments concerning the problems of objective-risks (section 2.2). There we said that a decision rule depending on the expected values z^{Γ} only prevents a consideration of the objective-risks and, therefore, supposes the decision maker to be indifferent against objective-risks. Certainly, this is not reasonable in the general case; hence the developed linear model is only allowed to be a special case. But, which of our assumptions excludes the consideration of objective-risks ? Preserving practicability we are enforced to guarantee the existence of a cardinal utility function meeting the expected-utility conditions; hence these conditions which refer to the preference relation on *M* must not be given up. Accordingly, we have to investigate whether the assumption (I) blocks the consideration of the objective-risks. That is indeed the case, for condition (I) means the following: let $\Omega=\{1,2\}$, $\mu(\{1\})=\alpha$ and $\mu(\{2\})=1-\alpha$ define a probability space. Let x, y be some vectors of R^n. The prospect $i_{\alpha x+(1-\alpha)y}$, i.e. "to get $\alpha x+(1-\alpha)y$ with certainty" may be written as a constant n-dimensional random variable $w : \Omega \rightarrow R^n$ with $w(\omega)=\alpha x+(1-\alpha)y$ for all $\omega\varepsilon\Omega$. The prospect $\alpha i_x+ + (1-\alpha)i_y$, i.e. "to get x with probability α and y with probability $1-\alpha$" may be written as follows: $v(1)=x$ and $v(2)=y$. Condition (I) therefore requires w as desirable as v. The matrix of intercorrelations of w is the unit-matrix, whereas this is certainly not the case for v, at least for some $x,y\varepsilon R^n$. But at the same time we have $E(w)=E(v)$ where E is the expectation operator. Hence (I) assumes indifference against objective-risks.

Result: 1) If distinctive behavior against objective-risks is to be considered, condition (I) of theorem 3.1 must be suppressed. Without (I) there is nothing said on the pattern of the utility function except for continuity, monotoneity and integrability. Choosing a special function requires a special argument. Choosing the linear function has been based on the assumption (I). If it is reasonable to assume indifference against objective-risks we may refer to section 2.4.4. for a determination of the correct weights a_i.

2) Assuming indifference against objective-risks together with the expected-utility theorem imply condition (I) and therefore the linear structure of the preferences on Im z^Γ, the space of vectors of expected values, and thus a special form of compromising conflicting objectives. But this linear structure is not plausible in every case; hence, if we want to treat indifference against objective-risks without prejudicing a certain kind of compromise-making, we will use another approach: by a *superobjective under indifference against objective-risks* for the decision situation Γ we mean an objective (H,Q,g) under certainty for the class $\{\Gamma\}$ with $H = \{\gamma\}$ where γ is the equivalence relation induced by the vector u^Γ of the isolated utility expectations: the Bernoulli-principle is applied to the isolated partial objectives. Q may be understood as a preference relation on Im u^Γ and g may be understood as a function of Im u^Γ in R. Additionally we require Q to meet the following conditions:

(a) Q is weakly monotone with respect to $\leq$.

(b) Q is continuous.

As before it follows that g is continuous and monotone increasing; naturally, a theorem of the expected-utility type for g does not hold: assuming (I) but allowing for other ways of comprise-making than the linear one the property of expected utilities can no longer be required. In this case alone we have a formal identity of the vector maximum problem and the problem of multi-objective decision making under uncertainty (the linear case turns out to be a special case of this type). We may therefore refer to our excursus on the vector maximum problem.

3.2. *The Quadratic Model*

In order to allow for the phenomena of objective-risks and to analyse them we will develop a model similar to the objective functions of portfolio selection theory. For this purpose we proceed in a way that is analogous to the approach to the one-dimensional case given by Farrar [1962].

Let I^n be a convex subset of R^n containing inner points. Assume that $u : I^n \rightarrow R$ is a utility function unique up to positive linear transformations. Let x^o be an inner point of I^n; suppose that u has continuous partial derivatives of the second order at all inner points of I^n. The Taylor-expansion of u with respect to x^o is written in the following way, if only terms up to the second order are considered:

$$u(x) = u(x^o) + \sum_{i=1}^{n} a_i(x_i - x_i^o) + \sum_{i=1}^{n} \sum_{j=1}^{n} b_{ij}(x_i - x_i^o)(x_j - x_j^o)$$

where $a_i = \partial_i u(x^o)$ and $b_{ij} = \partial_i \partial_j u(x^o)$ $(i,j=1,...,n)$. Since u is monotone increasing with respect to all variables we have $a_i > 0$ for all $i=1,..,n$. The coefficients b_{ij} reflect the specific behavior against the objective -risks; this is seen as follows: let μ be a probability measure on I^n; to facilitate the discussion we set $x^o = \int_{I^n} x \, d$, i.e. x^o equals the vector of expected values of the partial objectives. We assume that x^o is an inner point of I^n; then we have

$$\text{(A)} \qquad U(\mu) = \int_{I^n} u \, d\mu = u(x^o) + \sum_{i=1}^{n} \sum_{j=1}^{n} b_{ij} \text{cov}(x_i, x_j)$$

(Here x_i is understood as a random variable). b_{ij} depends on x^o. Thus the preference functional has two components only: a component directly depending on the expected values of the partial objectives and another component explicitly evaluating the objective-risks by means of the parameters $b_{ij}=b_{ji}$; these parameters are locally determined only.

Other than in the one-dimensional case one cannot speak of risk-aversion or risk-love without any specification[1], for the risk here is not represented by a real number but by a matrix of real numbers: the local behavior against the risk is more distinctive in the multi-dimensional case. Hence we may define:

A decision maker is *partially an objective-risk-averter with respect to the criteria i and j at the criteria-vector* $\bar{x}$, if

$$\partial_i \partial_j u(\bar{x}) < 0.$$

In the same way an objective-risk-lover and an indifferent behavior against objective-risk are defined. If a decision maker is indifferent against objective-risk with respect to all pairs (i,j) and at all points $\bar{x}$, he is indifferent against objective-risks altogether: if for all $\bar{x}$ and all $i,j=1,\ldots,n$ we have $\partial_i \partial_j u(\bar{x}) = 0$, then u is a linear function and we have the model of the preceding section.

In order to avoid that the parameters b_{ij} depend on the probability measure concerned, we assume u to be quadratic:

$$u(x) = c + \sum_{i=1}^{n} a_i x_i + \sum_{i=1}^{n} \sum_{j=1}^{n} b_{ij} x_i x_j .$$

For all i we have $a_i > 0$. Let μ be a probability measure on I^n:

$$U(\mu) = \int_{I^n} u \, d\mu = c + a.x^o + \sum_{i,j=1}^{n} b_{ij}(\mathrm{cov}(x_i,x_j) + x_i^o x_j^o)$$

$$\text{(A')} \qquad = u(x^o) + \sum_{i,j=1}^{n} b_{ij}\mathrm{cov}(x_i,x_j)$$

according to our previously given formula.

Considering positive linear transformations $t_i x_i + v_i$ for all $i=1,\ldots,n$ we define

(i) $\bar{b}_{ij} = b_{ij}(t_i t_j)^{-1}$ (ii) $\bar{a}_i = a_i/t_i - 2\sum_{j=1}^{n} \bar{b}_{ij} v_j$

(iii) $\bar{c} = c - \sum_i \bar{a}_i v_i - \sum_{i,j} \bar{b}_{ij} v_i v_j$

1) Arrow [1965, p. 30]

$$\bar{u}(y) := \bar{c} + \sum_{i=1}^{n} \bar{a}_i y_i + \sum_{i,j=1}^{n} \bar{b}_{ij} y_i y_j$$

Then we have $\bar{u}(y) = u(x)$ if $y_i = t_i x_i + v_i$ holds. As formula (A') shows we have $\int \bar{u}\, d\bar{\mu} = \int u\, d\mu$ if $\bar{\mu}$ is the analogon of μ generated by the transformation above described, i.e. replacing u by $\bar{u}$ corresponds to a substitution cardinal utility function on I^n by the analogous function on $\bar{I}^n$. The formulae (i) through (iii) represent transformation rules for replacing one representation of the system of objectives by another; thus quadratic preferences remain quadratic, a result of the type related to the linear and hyperbolic cases.

Formulae (A) and (A') give rise to another interesting interpretation: under uncertainty the problem of multi-objective decision making obviously has two dimensions; on the one hand the objectives have to be balanced with respect to their content: this balancing is represented by the term $u(x^o)$, the utility of the vector of expected values of the partial objectives. On the other hand the risk between the partial objectives has to be sized up: this judgment is expressed by the covariance term representing the distinctive local evaluation of the covariances between the partial objectives. But at the same time we see that these two dimensions cannot be isolated, since the function u involves a certain kind of risk-judgment. This has particularly become clear by treating the linear model: assuming indifference against objective-risks enforces a very specific kind of balancing of the objectives with respect to their content. For that reason, in that case the superobjective has been defined differently from the general case in order to admit compromising methods other than the linear one. The over-all Bernoulli principle has had to be weakened to an ordinal rational principle. For sake of a preservation of the operationality such a weakening seems not possible in the general case: it is very difficult to determine a preference relation immediately on the set of probability measures.

If I^n is interpreted as the closed convex hull of $\text{Im } Z \cap R^n$, the quadratic utility function u may be understood as a superobjective function analogous to the procedure of the preceding section. The parameters a_i and b_{ij} $(i,j=1,\dots,n)$ can be determined in the following way:

1. The decision maker states which covariances are to influence the superobjective positively and which negatively (partial objective-risk lover or averter). The covariances are provided with a sign accordingly so that all terms have a positive influence on the superobjective function.

2. The generated expressions are interpreted as partial objectives and the parameters are determined by the procedure of determining linear weights. In so doing the parameters b_{ij} are derived up to a joint positive multiplier $a_{n+1} > 0$.

3. It remains to solve the linear problem

$$u(x) - c = \sum_{i=1}^{n+1} a_i x_i$$

where x_{n+1} is some real variable to be interpreted as a risk-objective; its dimension naturally depends on the dimension of x and of the parameters b_{ij} determined before. This problem is solved in the previous-mentioned way.

Since u itself is unique only up to positive linear transformations the objective function u or U respectively is therefore found.

3.3. *The Hyperbolic Model*

3.3.1. *An Axiomatical Treatment of Hyperbolic Preferences*

Theorem 3.2.[1] Let $\preccurlyeq$ be a complete continuous preference relation on R^n_+ with the following properties:

(i) $\preccurlyeq$ is monotone on $I^n_+ := \{y \varepsilon R^n | 0 << y\}$. For all $x, y \varepsilon R^n_+$ with $x+y >> 0$ and $x \preccurlyeq y$ and all $\alpha \varepsilon >0,1<$ we have $x \prec \alpha x + (1-\alpha)y$ (hence $\preccurlyeq$ is strictly convex on I^n_+).

(ii) If $y \varepsilon R^n_+$ and $y \notin I^n_+$ hold, we have $y \sim 0$.

(iii) Let $\preccurlyeq$ be homogeneous in the following sense: for each positive dilatation t of R^n $\preccurlyeq$ and $\preccurlyeq_t$ are identical if we define

$$tx \preccurlyeq_t ty \iff x \preccurlyeq y.$$

Then $\preccurlyeq$ is a hyperbolic preference relation on R^n_+.

Proof: The set $H := \{y \varepsilon R^n_+ | \sum_{i=1}^{n} y_i = 1\}$ is compact. Hence the preference relation has a greatest element on H (theorem in the appendix) uniquely

1) For another axiomatical treatment we refer to Harsanyi [1959, pp. 325-331]. Treating Nash's solution for the cooperative n-person game Harsanyi introduces two assumptions immediately implying (iii) in place of (iii); he is led to the special case of mutually identical exponential weights. For another use of the Cobb-Douglas utility function we refer to Miller and Starr [1960].

determined by (i); by (ii) none of the coordinates of this element equals 0. Since $\preccurlyeq$ is monotone this element is at the same time the greatest element of $\bar{H} = \eta(H)$, it is denoted by x. For each $y \varepsilon R_+^n$ we write

$$u_x(y) = \prod_{i=1}^{n} y_i^{x_i}$$

It easily seen that u_x has its uniquely determined maximum with respect to $\bar{H}$ in x (proposition 2.5.).

Let F be some indifference class of $\preccurlyeq$ and $y \varepsilon F$. We define $t_i = y_i/x_i$ for all i=1,...,n and denote by t the mapping defined by $t(w) = t.w$. Then y is the greatest element of tH, since $tx = y$ and $tw \preccurlyeq tx$ hold by (iii) because we have $w \preccurlyeq x$ for all $w \varepsilon H$. Additionally we have $u_x(tw) = u_x(t)u_x(w)$. Hence u_x has its maximum at y with respect to tH. Thus at each point $y \varepsilon F$ the set tH is the tangent hyperplane to the hypersurface given by $u_x(w) = u_x(y)$. If for some $\bar{y} \varepsilon F$ we had $u_x(\bar{y}) > u_x(y)$ then the hypersurface given by $u_x(w) = u_x(\bar{y})$ must intersect the indifference class F. The hyperplane tangent to that hypersurface at a point of this intersection must intersect F too, and the tangent point cannot be the greatest point with respect to $\preccurlyeq$. In the same way the case $u_x(\bar{y}) < u_x(y)$ is brought to a contradiction. Hence we have $u_x(\bar{y}) = u_x(y)$ which completes the proof. ¶

The exponential weights of the utility function can be determined by the procedure set forth in 2.4.5. up to a joint positive multiplier.

Analysis of the Assumptions

(i) Strict convexity reflects a decision behavior searching for real compromises: each mixture of two extreme likely treated alternatives is strictly preferred to both alternatives. Such behavior may be appreciated as rational.

(ii) This assumption limits the applicability of the model to partial objectives with a natural origin. If such a criterion belongs to a system of objectives, i.e. if it is really pursued, it is reasonable to ask for positive achievements of this criterion, for otherwise this partial objective would ex ante be treated as inferior. Dual criteria are examples with natural origins.

(iii) This assumption is a further limitation of the applicability of the considered model: it means that the numerical value of the rates of substitution do not depend on the units of measurement of the partial objectives. Whether this assumption is reasonable remains to be examined in each special case.

3.3.2. *On the Uniqueness of the Utility Function*

As theorem 3.2. points out the utility function $u_a(x)$ is unique only up to continuous monotone increasing functions in R, since $u_a(x)$ has been derived only on the basis of indifference classes. Assuming indifference against objective-risks R^n_+ may be understood as the set of imaginable combinations of expected values provided with a hyperbolic preference ordering given by the decision maker. Thus this ordering is only to balance the conflicting objectives in the proper sense and not to deal with the problems of objective-risks. In this case it is all the same whether u_a or $\psi \cdot u_a$ is chosen as objective function, as long as ψ is a continuous increasing function in R, i.e. the ordinal utility concept is sufficient here.

But as soon as allowance is to be made for objective-risks, R^n_+ must be thought of as the set of imaginable realizations of criteria-vectors, and the set of probability measures on R^n_+ must be considered. Naturally, there exist several preference orderings on this latter set whose restriction to R^n_+ coincides with the relation $\preccurlyeq$ of theorem 3.2 . Different cardinal utility functions correspond to these orderings, each of them representing the same *ordinal* function on R^n_+. The problem is now to find the true monotone transformation ψ allowing the adequate treatment of objective-risks. Again, it becomes clear that the problem of multi-objective decision making under uncertainty comprises two components: the balancing problem in the proper sense is solved by the function u_a; treating the objective-risks is ensured by the monotone transformation ψ. The problem of finding the true monotone transformation ψ coincides with the problem of finding one-dimensional risk-preference functions; this problem is here supposed to be solved.

Nevertheless we will solve this problem for a particular case which is important for our purposes. Even assuming the cardinal utility function itself to be of the hyperbolic type involves the problem of finding the correct vector a of exponential weights, since our procedure determines a only up to positive multipliers; but replacing u_a by u_{ka} with $k > 0$ is not feasible in the cardinal sense:

$$u_{ka} = (u_a)^k.$$

The 'true' value of k can easily determined as follows: choosing some $x, y \varepsilon I^n_+$ with $x < y$, a real number $\alpha \varepsilon]0,1[$ is determined such that

$$i_x \sim (1-\alpha) i_y$$

holds. That can easily be done by a simple process of interval partitioning and obtaining suitable information from the decision maker. For the true k we have the following condition:

$$(u_a(x))^k = (1-\alpha)(u_a(y))^k$$

or

$$k = \frac{\log(1-\alpha)}{\log u_a(x) - \log u_a(y)}$$

If we may take $y=(1,1,\ldots,1)$ the formula is simplified to

$$k = \log(1-\alpha)/\log u_a(x)$$

By this, on principle, the problem is solved.

Remark: Considering u_a as a cardinal utility function this model is in some sense a polar pendant to the linear model: the linear model does not allow for covariances at all, the quadratic model deals just with all covariances, i.e. with the joint stochastic variations of *two* criteria at a time, the hyperbolic model considers the joint variations of all n criteria simultaneously. A Taylor-expansion which is to take account of these effects has to comprise terms up to the n-th order.

3.4. *A Collection of Models*

1) *Indifference against objective-risks* means that actions a and b are not treated differently as long as they are provided with the same vector of certainty equivalents separately formed for each partial objective, e.g. as

$$z^\Gamma(a) = \int_{Y_{\Gamma,a}} Z^\Gamma(a,.)\, d\mu_a = \int_{Y_{\Gamma,b}} Z^\Gamma(b,.)\, d\mu_b = z^\Gamma(b)$$

holds.

If, in addition, the model of a Bernoulli-utility function is assumed, we necessarily come up with the

a) *linear model*

$$u(a,\omega) = \sum_{i=1}^{n} \alpha_i Z_i^\Gamma(a,\omega)$$

$$U(a) = \sum_{i=1}^{n} \alpha_i \int Z_i^{\Gamma}(a,.)\, d\mu_a = \sum_{i=1}^{n} \alpha_i z_i^{\Gamma}(a)$$

where $\alpha_i > 0$ for all i and $a\varepsilon A$ and $\omega\varepsilon Y_{\Gamma,a}$. The weights α_i can be determined by the method set forth in 2.4.4.

If the over-all Bernoulli model is given up we obtain the general

b) *vector maximum problem* based on the utility expectations of the partial objectives which are separately formed: we assume monotone, continuous, differentiable, and convex preferences on the set of vectors of utility expectations; under additional assumptions on this set its greatest element can be determined by the method depicted in 2.4.7. Naturally the linear model may be treated as a special case of b).

2) *Consideration of objective-risks*

If the decision maker does not behave indifferently against objective-risks we generally choose the solution model of a superordinate utility function with the expected utility property. This has the advantage that operational models can be derived, which cannot be guaranteed with an ordinal concept on the set of probability distributions of results.

The disadvantage is that the compromising behavior and the bahavior against the objective-risks cannot be varied independently. Such a variability could be obtained in the case of indifference against objective-risks by introducing an ordinal utility concept without destroying the operationality of the chosen method. Applying the Bernoulli-principle links up certain forms of behavior against objective-risks with certain forms of compromising behavior, but not in a one-to-one way; by monotone transformations of a utility function on the set of consequences different structures of behavior against objective-risks can be generated (but not arbitrary ones) without changing the compromising behavior.

a) *The quadratic model*

$$u(a,\omega) = c + \alpha.Z^{\Gamma}(a,\omega) + Z^{\Gamma}(a,\omega)^t.B.Z^{\Gamma}(a,\omega),$$

where $c\varepsilon R$, $\alpha\varepsilon R_+^n$ and B is a symmetric (n,n)-matrix.

u represents a certain compromising behavior, each monotone transformation of u generates the same compromising behavior but, possibly, different forms of behavior against objective-risks. Considering the proper quadratic model we assume u to have the expected-utility property in the indicated form, i.e.

$$U(a) = \int u(a,.)\, d\mu_a = c + \alpha . z^\Gamma(a) + z^\Gamma(a)^t . B . z^\Gamma(a) +$$

$$+ \sum_{i,j=1}^{n} b_{ij} \operatorname{cov}(Z_i^\Gamma(a,.), Z_j^\Gamma(a,.))$$

The determination of the parameters c,α,B can be carried out along 3.2.. B includes a partial evaluation of objective-risks.

b) *The hyperbolic model*

Partial objectives provided with a natural zero point of measurement can be treated by the hyperbolic model in order to form a compromise:

$$u(a,\omega) = \prod_{i=1}^{n} (Z_i^\Gamma(a,\omega))^{\alpha_i} \qquad \alpha_i > 0.$$

If u is to have the expected-utility property the vector α must be normalized such that the true behavior against objective-risks is represented. This normalization can be done by using the procedure described in 3.3.2. The vector α can be determined up to positive multipliers by the method of 2.4.5.

c) *The general model with (INC)*

The linear as well as the hyperbolic models are special cases of preference structures satisfying the condition (INC) of 2.4.4. The general model is given by

$$u(a,\omega) = \sum_{i=1}^{n} \alpha_i f_i(Z_i^\Gamma(a,\omega))$$

where $\alpha_i > 0$ and f_i is monotone increasing and continuous for all i. Again, by u is determined the compromising behavior only, an additional monotone transformation of u adjusts the behavior against objective-risks. The weights α_i are determined by 2.4.6.

From the above-mentioned one the linear model follows by $f_i := \text{id}$ for all i, the hyperbolic model can be derived by taking $f_i := \log$ for all i and a monotone transformation of u by $u \rightarrow e^u$.

Remark: Naturally, each of the described models may apply to the general vector maximum problem by some simple replacing:

$$u(a,\omega) \quad \text{by} \quad U(a)$$

$$Z_i^\Gamma(a,\omega) \quad \text{by} \quad z_i^\Gamma(a).$$

4. *Application*

In this section some results set forth in the preceding chapters will be applied to a case of management planning. To this end we will investigate problems of planning the firm's product range (quality and extent) for one time period in connection with simple problems of factor procurement, of sales and of financing which are associated with the planning of the product range. We will assume given capacities not to be extended and will take in factor markets and markets for the sale of the firm's products by certain very simple conditions. For sake of a lucid exposition we do not treat several periods by a flexible planning approach: we will formulate a one-period stiff planning model. We will investigate a firm already existing at the start of the period and which intends to exist after the end of the period concerned. In order to simplify our arguments we assume that virtually only one sort of raw material is employed as an input factor which can be processed to n different products by a single-stage production.[1] Thus the problem of planning the product range involves choosing the sort of product to be produced amongst the n technically possible ones, and simultaneously, fixing the respective amounts to be produced.[2] Choosing the alternatives is determined by the states of the factor markets and of the markets for the sale of the firm's goods, by the financing opportunities, by the possible techniques of production and by the capacities of the firm's plants and of the factory labour availability.

1) Factor procurement

The firm can agree on different contracts for the delivery of raw material; activity parameters have the whole quantity M to deliver during the period and the number m of contract dates for partial deliveries of the same amount M/m. The first contract date is the begin of the period. The price per unit of M depends on M,m and an index i corresponding to the line of business of the suppliers: q(M,m,i).

The transactions can be financed by credit allowed by the supplier. But credits and interest agreed upon have to be repaid before the next contract date.

---

1) One might think of the production of different plastic goods consisting of the same material. This single input factor may also be understood as a bundle of limitational factors which are strictly joint on the factor market.
2) Cf. Gutenberg [1968, Vol. I, p. 155]

2) Inventory of raw material

The firm owns a storage house of a maximal capacity K; in appropriate instances external storage-capacities can be hired at any amount. External storage is expensive and uncomfortable (organizational and transporting problems); the management wishes to avoid it as far as possible.

3) Production

We assume that the firm's realizable production processes can be formalized by the following production correspondance: input factors are

a) raw material r measured by tons (e.g.)
b) machine labour: there are k machines, each of them can produce each of the n different products; but during the planning period each machine is to produce one kind of products only ($k \geq n$). The machine labour of the i-th machine will be denoted by L_i (measured by hours ($i=1,\ldots,k$)).
c) Direct human labour: w_i hours with respect to the i-th machine. Remaining labour performed is assumed to be independent of the production.

If no dead-locks of operation occur, the input will be connected with the respective amounts of output $x_1,\ldots,x_n$ by the correspondance P:

$$P \subset R_+^{2k+n+1} .$$

An element $v\varepsilon P$ ($v=(r,L_1,\ldots,L_k,w_1,\ldots,w_k,x_1,\ldots,x_n)$) means that, deadlocks supposed not to happen, the bundle x of products can be processed by employing r tons of raw material running the i-th machine during L_i hours and employing direct human labour with respect to this machine during w_i hours ($i=1,\ldots,k$).

Planning the product range under certainty would simply mean to choose an element $v\varepsilon P$. But v belonging to P especially means that v requires capacities at hand only, i.e. there are enough workmen and raw material available in order to run the required machines during the required period of time. If stochastic elements have some influence, this can no longer be guaranteed, P is (ex ante) not uniquely determined: if machines become inoperative or workmen fall ill, the chosen v may become unrealizable. Hence the management can only have probabilistic expectations with regard to P (possibly depending on activities concerning other areas of the firm: factor procurement, financing).

Hence, planning the product range under uncertainty means to select a

$v \in R_+^{2k+n+1}$ with the hope that v will in fact belong to the realized P. If this turns out to be not the case, adjustment activities become necessary; to avoid the need for a flexible planning technique we assume the respective activities to be uniquely determined: in those cases as much as possible will be produced under the given circumstances. This assumption does not have any consequences, it just enables us to get along without flexible planning techniques. Cases where the chosen v does not belong to the realized P are unpleasant, since they necessarily lead to short-term adjustment activities and planning revisions.

4) Sales

Let the firm be faced with a stochastic price-to-sales function. The firm behaves as a price-taker (thus we assume that the firm does not influence this function by advertising efforts (e.g.)). The prices vary stochastically: for each output vector x there exists a n-dimensional random variable $p(x) : \Omega \rightarrow R^n$ assigning to each x the distribution of the corresponding attainable price vectors.

5) Financing

The transactions can be financed by credits allowed by the suppliers or by using overdraft facilities. For reasons of the firm's general policy the latter facilities should be used as rarely as possible. We assume the credits can be obtained at any amount: this assumption is not unreasonable with the problem here considered, since the needed amounts will be relatively small compared with credits for investment financing. We therefore do not handle the problem of scarce financial funds.

Let (Ω, A, μ) be the probability space of environmental responses to the firm's activities; we assume it to be independent of the firm's actions.

The price index mentioned in 1) is a real random variable defined on that space; hence $q(M,m,i)$ becomes a stochastic function.

In this model a decision consists of choosing

a) $M \in R_+$, b) $m \in N$, c) $v \in R_+^{2k+n+1}$, d) the financing opportunity to be used and the related amount to be borrowed for each date j/m $(j=1,\ldots,m-1)$ within the period concerned.

We assume that the produced goods are immediately sold and that the proceeds obtained are received in cash before the next date of a delivery of raw material.

P is a random object defined on (Ω, A, μ) depending on M and m. For a process $v \varepsilon R_{+}^{2k+n+1}$ chosen by a certain decision, we denote by v_{ω} the process which is realized if the elementary event ω occurs in accordance to the convention made in 3). We assume that for each v the set $\{\omega\varepsilon\Omega | v\varepsilon P(M,m)(\omega)\} = \{\omega\varepsilon\Omega | v=v_{\omega}\}$ is measurable such that it makes sense to speak of the probability of v being realized.

Given some m, for each partial period and each elementary event ω the related values of i,v and q are to be specified:

$i_j(\omega)$ price index at the start of (partial) period (j,j+1)

v_{ω}^{j} part of vector v_{ω} realized in period (j-1,j) if v was planned (naturally the v_{ω}^{j} sum up to v_{ω}).

$p_{j\omega}(x)$ price vector yielded at the end of the period (j-1,j) given the output x.

(We use j in place of j/m).

The *stock of raw material* at the start of period (j,j+1) is given as follows: (by E is meant a decision in the above-mentioned sense)

$$B_{j\omega}(E) = M(j+1)/m - \sum_{t=o}^{j} pr_1 v_{\omega}^{t} \qquad (j=1,\ldots,m-1)$$

Let the stock at j=o be B_o (E)=M/m; the stock at j=m is given by

$$B_{m\omega}(E) = M - \sum_{t=o}^{m} pr_1 v_{\omega}^{t}.$$

Cash payments at the start of period (j,j+1) (j=0,...,m-1)

- hire for external storage $M_j(E)$
- raw material immediately paid for $y_j(E).q(M,m,i_j(\omega)).M/m$ where $y_j(E)$ denotes the fraction of accounts payable for raw material immediately paid (here it is stiffly planned again; it would be more adequate to decide on this fraction subject to the occurrence of certain events, but this would lead us to a much more complicated flexible planning model).
- costs usually paid at the begin of the period (hire, insurance premia etc.) $K_{j\omega}(E)$

Cash payments at the end of period (j,j+1)

- repayment of accounts payable for raw material $(1+\alpha)(1-y_j(E)).$ $.q(M,m,i_j(\omega)).M/m$ where α denotes the interest rate (per partial period) for the credit allowed by the supplier
- repayment of overdrawal $(1+\beta)z_{j\omega}(E)$
- costs usually paid at the end of the period (wages) $\bar{K}_{j\omega}(E)$
- further cash payments independent of the decision E (interest on long-term debt etc.) $S_{j\omega}$

Since the end of the period (j,j+1) coincides with the begin of the period (j+1,j+2) we have cash payments at the end of the period (j-1,j) as follows (j=1,...,m):

$$A_{j\omega}(E) = M_{j\omega}(E) + y_j(E)q(M,m,i_j(\omega))M/m + K_{j\omega}(E) +$$

$$+ (1+\alpha)(1-y_{j-1}(E))q(M,m,i_{j-1}(\omega))M/m +$$

$$+ \bar{K}_{j-1\omega}(E) + (1+\beta)z_{j-1\omega}(E) + S_{j-1\omega}.$$

At the begin of the planning period we have

$$A_{o\omega}(E) = M_{o\omega}(E) + y_o(E)q(M,m,i_o(\omega))M/m + K_{o\omega}(E)$$

Cash receipts at the begin of period (j,j+1)

- overdrawal $z_{j\omega}(E)$

Cash receipts at the end of period (j,j+1)

- sales revenue $p_{j+1\omega}(x_{j+1\omega}(E)).x_{j+1\omega}(E)$ where by $x_{j+1\omega}(E)$ is denoted the output produced according to the decision E.

Hence we have cash receipts at the end of the period of (j-1,j) = start of period (j,j+1) as follows:

$$E_{j\omega}(E) = z_{j\omega}(E) + p_{j\omega}(x_{j\omega}(E)).x_{j\omega}(E) \qquad (j=1,...,m)$$

and at the start of the planning period

$$E_{o\omega}(E) = z_{o\omega}(E)$$

We assume an initial stock H_O of liquid assets. The stock of liquid assets for the different partial periods are therefore as follows:

$$H_{j\omega}(E) = H_{j-1\omega}(E) + E_{j\omega}(E) - A_{j\omega}(E) \qquad (j=1,\ldots,m)$$

(liquid assets are supposed to be held in cash).

Since the overdraft facility is to be used as rarely as possible, $z_{j\omega}(E)$ is uniquely determined by solving the following sequence of problems:

$$\min \{H_{j\omega}(E) \mid H_{j\omega}(E) \geq 0 \text{ and } z_{j\omega}(E) \geq 0\} \qquad (j=0,\ldots,m)$$

The overdraft facility is to be used only as far as is necessary to ensure the firm's liquidity.

The criteria

a) The firm wishes to be enforced as rarely as possible to hire external storage capacities; but there are no time preferences with respect to this inconvenience:

$$\Omega_2(E) = \{\omega\varepsilon\Omega \mid \text{for all } j=0,\ldots,m \text{ we have } B_{j\omega}(E) \leq K\}$$

is the event when, given the decision E, the storage capacity is sufficient in order to store the amount of raw material to be delivered at another time during all of the planning period. For an arbitrary subset $\bar{\Omega}$ of Ω we define

$$\chi_{\bar{\Omega}}(\omega) = \begin{cases} 1 \text{ if } \omega\varepsilon\bar{\Omega} \\ 0 \text{ elsewhere} \end{cases}$$

to be the *characteristic function of* $\bar{\Omega}$.

Then, as a first criterion we may define the following function

$$Z_2(E,.) = \chi_{\Omega_2(E)}(.)$$

b) The firm wishes to reorganize the production process as rarely as possible:

$$\Omega_3(E) = \{\omega\varepsilon\Omega \mid v\varepsilon P(M,m)(\omega)\}$$

is the event when the production process runs as planned.

$$Z_3(E,.) = \chi_{\Omega_3(E)}(.)$$

c) Using the overdraft facility should be as rare as possible; here again no time preferences exist:

$$\Omega_4(E) = \{\omega\varepsilon\Omega \mid \text{ for all } j=0,\ldots,m \text{ we have } z_{j\omega}(E)=0 \text{ and } H_{m\omega}(E)\geq 0\}$$

is the event when the liquid assets at hand in addition to the credit allowed by the suppliers are sufficient to pay the agreed payments at each date and to guarantee a positive amount of liquid assets at the end of the planning period after having repaid all liabilities.

$$Z_4(E,.) = \chi_{\Omega_4(E)}(.)$$

d) Let $w_\omega(B_{m\omega}(E))$ be the value of the final stock of raw material. Depending on the production v_ω we have a depreciation $C_\omega(E)$ of the firm's plants. The profit may be defined by

$$G_\omega(E) = H_{m\omega}(E) + w_\omega(B_{m\omega}(E)) - C_\omega(E) - H_{o\omega}.$$

It equals the excess of liquid assets at the end of the planning period over those provided at the begin of the period, enlarged by the value of the remaining raw material and reduced by the depreciation generated by the production process. Valuation problems possibly arising here will not be treated.

$$Z_1(E,\omega) = G_\omega(E)$$

We define a decision E to be an element of $A := N\times R_+^{2k+n+1}\times\langle 0,1\rangle^n$: $E = (m,M,v,y)$ where the following notation is used:

m	the number of partial deliveries
M	the total amount of raw material to be supplied
v	the planned production vector
y	the vector of fractions of accounts payable for raw material immediately paid

By the set A the set of imaginable alternatives is described. By $Z_1,\ldots,Z_4$ a complete system of objectives is described. As previously we denote

$$z_i(E) = \int_\Omega Z_i(E,.)\,d\mu \qquad (i=1,\ldots,4)$$

Solution approaches

The classical "solution" to this problem is given by the chance-constrained programming model: the decision maker is assumed to be able to give lower bounds for achieving the criteria 2, 3, and 4. Then the problem is reduced to the following optimization problem:

$$\max \{z_1(E) | E \varepsilon A \text{ and } z_i(E) \geq \alpha_i \text{ for } i=2,3,4\}$$

where the parameters α_i are provided by the decision maker. On this approach it is necessary to remark on the following:

a) This approach supposes that a "solution" of the stated problem can be found based on the functions $z_1,\dots,z_4$ only; this assumption turns out to mean indifference against objective-risks.

b) Even assuming such indifference this approach leaves the decision maker with the problem of choosing the parameters α_i. We have previously emphasized that there are no satisfying reasons for assuming an a priori existence of those bounds. At most the determination of such bounds may be the result of a process during which the decision maker compares the effects of different parameter constellations with each other; consequently these parameters are no longer bounds. In general it is a complicated job to initiate the mentioned process at all. We have suggested an algorithm to organize it. But to solve the general problem of multiobjective decision making under uncertainty we are referred to the use of models fitting parameterized preference structures to the specific decision making behavior of the person concerned.

1) The quadratic model

The quadratic model can deal with both indifference against objective-risks and consideration of objective-risks. We have the following superobjective function:

$$U(E) = \sum_{i=1}^{4} a_i z_i(E) + \sum_{i,j=1}^{4} b_{ij} z_i(E) z_j(E) + \sum_{i,j=1}^{4} b_{ij} \operatorname{cov}(Z_i(E,.), Z_j(E,.))$$

We have $\operatorname{cov}(Z_1(E,.),Z_j(E,.)) = \int_{\Omega_j(E)} Z_1(E,.)\, d\mu - z_1(E).\mu(\Omega_j(E))$ for all $j=2,3,4$. This expression equals zero if, on the average, the profit obtained in $\Omega \setminus \Omega_j(E)$ (i.e. for j=2: if storage capacity is not sufficient) equals the profit obtained in $\Omega_j(E)$ (i.e. if storage capacity is sufficient). Hence, if hiring external storage capacities strongly influences profit this expression will be noticeably different from zero. The given covariances therefore measure how strongly the criterion Z_j

influences the profit.

The covariances between the criteria Z_2, Z_3, Z_4 allow the following interpretation:

$$\mathrm{cov}(Z_i(E,.), Z_j(E,.)) = \mu(\Omega_i(E) \cap \Omega_j(E)) - \mu(\Omega_i(E)).\mu(\Omega_j(E))$$

These expressions clearly equal zero, if and only, if the related events are independent of each other; in general this is not the case: the event "the storage capacity is sufficient" depends on whether the production process runs as planned or enlarged stocks of raw material occur because of unexpected stand-stills of production.

If all b_{ij} vanish we have the case of indifference against objective-risks: the quadratic model becomes linear. In this case the optimal values of z_2, z_3, z_4 depend only on the subjective rates of substitution between the expected values of the criteria. These values may be interpreted as the "correct" parameters for a chance-constrained programming approach. By the way, this is true for every other treatment of the indifference against objective-risks.

If not all b_{ij} equal zero the optimal decision cannot even on principle be determined by a chance-constrained programming approach, there are no "correct" parameters α_i; as is easily seen the optimal decision need not be efficient with respect to the objective functions $z_1, \ldots, z_4$.

2) Hyperbolic models

Assuming the decision makers preference relation to satisfy condition (INC) enables us to consider the criteria by groups and to construct partial superobjective functions.

Case a) Indifference against objective-risks

The hyperbolic model is adequate for the group z_2, z_3, z_4 because of their specific structure (natural zero and unit of measurement). Naturally, we are no longer within a cardinal utility concept. The function $u_a(z_2, z_3, z_4)$ may therefore be understood as a partial superobjective function of the Cobb-Douglas type whose parameter a indicates the relative importance of the criteria, its determination is as previously mentioned. Balancing the functions u_a and z_1 is a 2-dimensional problem to be solved by the depicted algorithm.

With this model in mind we see that even assuming indifference against objective-risks does not make the chance-constrained programming approach reasonable because of the complexity of all the internal interdependencies of the different criteria. Again the assumption of an a priori existence of the depicted parameters cannot be accepted.

Case b) Consideration of objective-risks

Applying the hyperbolic model to the group Z_2, Z_3, Z_4 leads to a very simple result:

$$u_a(E,\omega) = \begin{cases} 1 & \text{if } \omega \varepsilon \bigcap_{i=2}^{4} \Omega_i(E) \\ 0 & \text{otherwise.} \end{cases}$$

Hence the partial superobjective function may be understood as the propability of all "side conditions" being met. The weights have no meaning unless they equal zero for some i.

In order to balance Z_1 and u_a an arbitrary model of the depicted kind may be used.

3) Summary

The given theoretical analysis and application can be summarized as follows:

a) Assuming indifference against objective-risks we are led to a vector maximum problem; using cardinal utility theory in solving this problem we get linear preferences. Within an ordinal concept this case can be treated by quadratic, hyperbolic or other models which can be fitted to the decision maker's specific behavior. On the other hand, it can be solved by the algorithm set forth in 2.4.7. if the conditions are met.

In either case we get a vector of expected values of the partial objectives which is efficient with respect to the set of feasible vectors of this kind.

b) If objective-risks are allowed for we have in general no longer a vector maximum problem: we have shown that objective-risks play an important role . It follows that the resulting optimal decision need not be efficient in the sense of a).

Since in this case there is no approach we can use to treat every preference structure, special structures are to be fitted to the decision maker: we have characterized a class of utility functions which can be trimmed along the same line, and have, analogously to portfolio selection theory, developed a quadratic utility function which can deal with the effects of objective-risks in a simple manner.

Which of these approaches is to be used depends on the decision maker and on the decision situation concerned. In appropriate instances it is advisable to solve the same problem with different tools and to compare the respective results.

Appendix

Theorem A: Let X be a non-empty compact topological space.[1] Let $\preccurlyeq$ be a complete preference relation on X. Let $\preccurlyeq$ be continuous with respect to the topology of X, i.e. for each $x \varepsilon X$ the sets $\{y \varepsilon X | x \preccurlyeq y\}$ and $\{y \varepsilon X | y \preccurlyeq x\}$ are closed. Then there exists a greatest element of X.

Proof: For each $x \varepsilon X$ we define a set $F_x := \{y \varepsilon X | x \preccurlyeq y\}$. The family F of all such sets forms a filter-base[2] on X; this is obvious. Since X is compact the filter generated by F has an adherent point[3], i.e. $z \varepsilon \bar{F}_x = F_x$ for all $x \varepsilon X$. Hence for all $x \varepsilon X$ we have $x \preccurlyeq z$ which completes the proof. ¶

Corollary: Let A and $E \neq \emptyset$ be topological spaces. We assume on E a complete continuous preference relation $\preccurlyeq$. Let B be a quasi-compact subspace of A and $e : A \rightarrow E$ be continuous. Then there is a $x \varepsilon B$ such that for all $y \varepsilon B$ we have $e(y) \preccurlyeq e(x)$.

In order to prove this we give the following

Lemma: Let E be a topological space and $\preccurlyeq$ a continuous complete preference relation on E. Then $\bar{E} = E|_{\sim}$ is completely ordered and separate, and the canonical projection $p : E \rightarrow \bar{E}$ is continuous and an order-faithful homomorphism.

Proof of the Lemma: Suppose $p(x), p(y)\ \bar{E}$ and $p(x) \neq p(y)$. Without any loss of generality we assume $x \prec y$. If $(]\leftarrow, x[,]y, \rightarrow[)$ is a Dedekind decomposition, both $]\leftarrow, x[$ and $]y, \rightarrow[$ are open sets and saturated with re-

1) Bourbaki [1965, chap. 1, § 9, No 1, Def. 1]
2) Bourbaki [1965, chap. 1, § 6, No 3, Def. 3]
3) Bourbaki [1965, chap. 1, § 9, No 1, Def. 1]

spect to ~ [1] and thus $p(>\leftarrow,x>)$ and $p(<y,\rightarrow<)$ are neighborhoods of $p(x)$ and $p(y)$ respectively satisfying the Hausdorff-axiom.[2] If there were a $z \in E$ with $x \prec z \prec y$, $p(>\leftarrow,z>)$ and $p(<z,\rightarrow<)$ would satisfy the Hausdorff-axiom. ¶

Proof of the corollary: $f = p \circ e$ is continuous. Hence Im f is quasi-compact in $\bar{E}$; since $\bar{E}$ is separate (lemma!), Im f is closed. The ordering of $\bar{E}$ is continuous and its restriction to Im f meets the conditions of theorem A. Hence the assertion is proved by using the fact that p is order-faithful. ¶

---

1) A set $M \subset E$ is saturated with respect to ~, if $x \in M$ $y \in E$ $x \sim y \Rightarrow y \in M$.
2) Bourbaki [1965, chap. 1, § 8, No 1, Prop. 1, Cond. (H)]

References

Albach, H.: Das Investitionsbudget bei Unsicherheit. Zeitschrift für Betriebswirtschaft 1967, pp. 503 - 518

Albach, H.: Informationswert. In: Handwörterbuch der Organisation. Hrsg.: E. Grochla. Stuttgart 1969, col. 720 - 727

Albach, H. and Schüler, W.: On a method of capital budgeting under uncertainty. Journal of Mathematical and Physical Sciences (India), Vol. 4, No. 3, 1970, pp. 208 - 226

Arrow, K.J.: Social choice and individual values. New York - London 1951

Arrow, K.J.: Aspects of the theory of risk-bearing. Helsinki 1965

Aubin, J.-P. and Näslund, B.: An exterior branching algorithm. Working-paper 72-42 European Institute for Advanced Studies in Management. Brüssel 1972

Aumann, R.J.: Subjective programming. In: Human judgements and optimality. Ed.: M.W. Shelly and G.L. Bryan. New York - London - Sydney 1964, pp. 217 - 242

Aumann, R.J.: A survey of cooperative games without side payments. In: Essays in mathematical economics. In honour of Oskar Morgenstern. Ed.: M. Shubik. Princeton 1967, pp. 3 - 27

Balderstone, F.E.: Optimal and feasible choice in organisations having multiple goals. Working Paper No. 12, Management Science Nucleus, Institute of Industrial Relations, University of California, Berkeley 4. California Febr. 1960

Belenson, S.M. and Kapur, K.C.: An algorithm for solving multicriterion linear programming problems with examples. Operational Research Quarterly, Vol. 24, No. 1, 1973, pp. 65 - 77

Benayoun, R., de Montgolfier, J., Tergny, J. and Laritchev, O.: Linear Programming with multiple objective functions: Step Method (STEM). Mathematical Programming, Vol. 1, No. 3, 1971, pp. 366 - 375

Bidlingmaier, J.: Zielkonflikte und Zielkompromisse im unternehmerischen Entscheidungsbereich. Wiesbaden 1968

Bourbaki, N.: Eléments de mathématique, Livre III, Topologie générale, Chap. 9. Paris 1958

Bourbaki, N.: Eléments de mathématique, Livre III, Topologie générale, Chap. 1 and Chap. 2. Paris 1965

Bourbaki, N.: Eléments de mathématique, Livre I, Théorie des ensembles. Paris 1970

Carnap, R. and Stegmüller, W.: Induktive Logik und Wahrscheinlichkeit. Wien 1959

Charnes, A. and Cooper, W.W.: Management models and industrial applications of linear programming. New York - London - Sydney 1967, Vol. I

Churchman, C.W. and Ackoff, R.L.: An approximate measure of value. Operations Research 2 (1954), p. 176

Churchman, E.W., Ackoff, R.L. and Arnoff, E.L.: Introduction to operations research. New York - London 1959

Debreu, G.: Theory of value. New York - London 1959

Dinkelbach, W.: Über einen Lösungsansatz zum Vektormaximumproblem. In: Unternehmensforschung heute. Ed.: M. Beckmann. Berlin - Heidelberg - New York 1971, pp. 1 - 13

Dinkelbach, W. and Isermann, H.: On decision making under multiple criteria and under incomplete information. In: Cochrane, J.L. and Zeleny, M. (Ed.), Multiple criteria decision making. Columbia/South Carolina 1973, p. 305

Dyer, J.S. An empirical investigation of a man-machine interactive approach to the solution of the multiple criteria problem. University of South Carolina Press 1972

Dyer, J.S.: A time-sharing computer program for the solution of the multiple criteria problem. Management Science, Vol. 19, No. 12, 1973, pp. 1379 - 1383

Eckenrode, R.T.: Weighting multiple criteria. Management Science 12 (1965), p. 180

Fandel, G.: Optimale Entscheidung unter mehrfacher Zielsetzung. Berlin - Heidelberg - New York 1972

Fandel, G. and Wilhelm, J.: Zur Entscheidungstheorie bei mehrfacher Zielsetzung. Mitteilungen aus dem Bankseminar der Rheinischen Friedrich-Wilhelms-Universität Nr. 6. Bonn 1974

Farrar, D.E.: The investment decision under uncertainty. Englewood Cliffs, N.J., 1962

Fishburn, P.: Utility theory for decision making. New York 197o

Fleming, M.: A cardinal concept of welfare. Quarterly Journal of Economics LXVI (1952), pp. 366 - 384

Geoffrion, A.M.: A parametric programming solution to the vector maximum problem, with applications to decisions under uncertainty. Stanford/California 1965

Geoffrion, A.M.: Resource allocation in decentralized non-market organizations with multiple objectives. Paper presented at the 2nd world congress of the Econometric Society. Cambridge (England) Sept. 1970

Gutenberg, E.: Grundlagen der Betriebswirtschaftslehre, Erster Band: Die Produktion. Berlin - Heidelberg - New York 1968

Harsanyi, J.C.: Cardinal welfare, individualistic ethics, and interpersonal comparisons of utility. The Journal of political economy Vol. 63 (1955), pp. 309 - 321

Harsanyi, J.C.: A bargaining model for the cooperative n-person-game. In: Contributions to the theory of games IV. Ed.: A.W. Tucker and R.D. Luce. Princeton 1959, pp. 325 - 355

Hart, A.G.: Risk, uncertainty, and the unprofitability of compounding probabilities. In: Studies in Mathematical Economics and Econometrics. Chicago 1942

Heinen, E.: Das Zielsystem der Unternehmung. Wiesbaden 1966

Ijiri, Y.: Management goals and accounting for control. Amsterdam 1965

Johnson, E.: Studies in multiobjective decision models. Lund 1968

Kaplan, A.D.H., Dirlam, J.B., Lanzilotti, R.F.: Pricing in big business. Washington 1958

Karlin, S.: Mathematical methods and theory in games, programming and economics, Vol. I. Reading (Mass.) - Palo Alto - London 1962

Keynes, J.M.: Treatise on probability. London - New York 1950

Klahr, D.: Decision making in a complex environment: the use of similarity judgements to predict preferences. Management Science Vol. XV 1969 (Theory), pp. 595 - 618

Knight, F.H.: Risk, uncertainty, and profit. Boston - New York 1921

Krelle, W.: Präferenz- und Entscheidungstheorie. Tübingen 1968

Krümmel, H.J.: Bankzinsen. Köln - Berlin - Bonn - München 1964

Krümmel, H.J.: Liquiditätssicherung im Bankwesen II. Kredit und Kapital 1969, pp. 60 - 110

Kuhn, H.W. and Tucker, A.W.: Nonlinear programming. In: Proceedings of the second Berkeley symposium on mathematical statistics and probability. Ed.: J. Neyman. Berkeley California 1951, pp. 481 - 492

Lang, S.: Algebra. New York 1965

Lenz, H.: Grundlagen der Elementarmathematik. Berlin 1961

Marglin, S.A.: Objectives of water-resource-development: A general statement. In: Design of water-resource systems. Ed.: A. Maass. Cambridge Mass. 1966

Markowitz, H.M.: Portfolio selection. New York - London 1959

Miller, D.W. and Starr, M.R.: Executive decisions and operations research. Englewood Cliffs 1960

Mosteller, P. and Nogee, P.: An experimental measurement of utility. The Journal of Political Economy 59 (1951), pp. 371 - 404

Näslund, B.: Decisions under risk. Stockholm 1967

v. Neumann, J. and Morgenstern, O.: Theory of games and economic behavior. Princeton 1946

Neveu, J.: Mathematische Grundlagen der Wahrscheinlichkeitstheorie. München 1969

Nievergelt, E.: Ein Beitrag zur Lösung von Entscheidungsproblemen mit mehrfacher Zielsetzung. Die Unternehmung, 25. Jg., No. 2, 1971, pp. 101 - 126

Pfanzagl, J.: Theory of measurement. Würzburg - Wien 1968

Pratt, J.W.: Risk aversion in the small and in the large. Econometrica 32 (1964), pp. 122 - 136

Rabusseau, R. and Reich, W.: Entscheidungen einer Person unter Unsicherheit. In: Proceedings in operations research (Vorträge der Jahrestagung 1971 DGU). Ed.: M. Henke, A. Jaeger, H.J. Zimmermann. Würzburg - Wien 1972, pp. 123 - 141

Radner, R.: Mathematical specification of goals for decision problems. In: Human judgements and optimality. Ed.: M.W. Shelly and G.L. Bryan. New York - London - Sydney 1964, pp. 178 - 216

Raia, A.P.: Goal-setting and self-control. The Journal of Management Studies 1965, p. 34

Riemenschnitter, A.: Die Kreditfinanzierung im Modell der flexiblen Planung. Berlin 1972

Rothenberg, J.F.: The measurement of social welfare. Englewood Cliffs, N.J., 1961

Sauermann, H. and Selten, R.: Anspruchsanpassungstheorie der Unternehmung. Zeitschrift für die gesamte Staatswissenschaft 118 (1962), pp. 577 - 597

Savage, L.J.: The foundations of statistics. New York 1954

Schmidt-Sudhoff, U.: Unternehmerziele und unternehmerisches Zielsystem. Wiesbaden 1967

Schneeweiß, H.: Entscheidungskriterien bei Risiko. Berlin - Heidelberg - New York 1967

Schubert, H.: Kategorien I. Berlin - Heidelberg - New York 1970

Simon, H.A.: Models of man. New York - London 1957

Theil, H.: On the symmetry approach to the committee decision problem. Management Science Vol. 9 (1963), pp. 380 - 393

Torgerson, W.S.: Theory and methods of scaling. New York 1958

Überla, K.: Faktorenanalyse. Berlin - Heidelberg - New York 1968

White, C.M.: Multiple goals in the theory of the firm. In: Linear programming and the theory of the firm. Ed.: K.E. Boulding and W.A. Spivey. New York 1960, p. 186

Wundt, W.: Willenshandlung und Wahlhandlung. In: Die Motivation menschlichen Handelns. Ed.: H. Thomae. Köln - Berlin 1965, p. 391

Zionts, S. and Wallenius, J.: An interactive programming method for solving the multiple criteria problem. Working-paper 74-10 European Institute for Advanced Studies in Management. Brüssel 1974